DATE DUE

NOV 04 '94			

HIGHSMITH 45-220

The Environmental Wars

The
Environmental
Wars

Reports from the Front Lines
David Day

St. Martin's Press   **New York**

Library of Congress Cataloging-in-Publication Data

Day, David,
        The Environmental Wars : Reports from the front lines /
    David Day.
                p.   cm.
            ISBN 0-312-04418-6
            1. Ecology.   I. Title.
    QH541.D39   1990
    363.7—dc20   89-77823
                                                    CIP

First published in Great Britain by Harrap Books Ltd.

First U.S. Edition

10   9   8   7   6   5   4   3   2   1

Contents

12 War of The Heart and Mind 293

Towards a negotiated peace settlement

Foreword

The Environmental Wars: is an encyclopaedia of ecological activism. It is an attempt – in a single volume – to take an overview of the movement in all its diverse concerns: from the fate of a rare egg to the survival of the planet. The book is arranged in such a way as to give the layman a grasp of the issues and implications of the ecology movement worldwide.

The Environmental Wars is a reference book or a compendium to be read selectively according to the interests and concerns of the reader. It may also be read as a continuous narrative. Its arrangement is such that it indeed progresses from the egg to the entire planet: from the black-market trade in rare eggs and endangered species to the prospect of the holocaust of nuclear war.

There may be some contention about the concept of the 'ecology wars' and the use of military language as a dramatic vehicle for reporting on what is largely a pacifist movement. However, I believe that the concept and the terminology accurately reflect the magnitude and seriousness of the conflict. In the final analysis, the outcome of the 'ecology wars' will prove more critical than any ever fought by the human race. What is at stake is not the dominance of one nation over another, but the survival of life on the planet.

Acknowledgments

I would like to thank the following individuals who at one stage or another over the past eight years helped me in the preparation of this book. However, this acknowledgment does not necessarily imply agreement with the views or opinions in this book. Furthermore, any errors or omissions are entirely my own responsibility.

Craig van Note, Allan Thornton, Dave Currey, Jenny Lonsdale, Jean-Paul Fortom-Gouin, Paul Watson, Jane Smith, Beverley Thorpe, Nick Carter, Cleveland Amory, Christine Stevens, Bill Jordan, Stefan Ormrod, John May, Tina Harper, Philip Cade, Campbell Plowden, Phillip Shabecoff, John Brookland, Geoffrey Lean, Lewis Registein, Terry Jones, Kathryn Court, Ray Gambell, Sidney Holt, Charles Secrett, Sarah Hambly, Jane McLernon, Richard Ellis, June Fenby, Catherine Caufield, Jeremy Cherfais, Patricia Forkan, Bob Hunter, Tom Garrett, Elke Stenzel, David Garrick, David McColl, Paul Vodden and Ian MacPhail.

In addition I would like to thank the following organizations:

Environmental Investigation Agency, Greenpeace, Sea Shepherd, Animal Welfare Institute, Fund For Animals, Humane Society of America, Royal Society For the Prevention of Cruelty to Animals, People's Trust For Endangered Species, Earth Trust, Cetacean Society, International Whaling Commission, Friends of the Earth, International Fund for Animal Welfare, Center for Environmental Education, Earth First! and The Beast.

Also a special thanks to Susanne McDadd, Derek Johns and the Canada Council.

Prologue:

The Human Pyramid

The Environmental Wars, a definition and body count

The Environmental Wars is dedicated to eight 'soldiers' – some would say martyrs – who have died violently in the front lines of the ecology wars. As this book demonstrates, there have been hundreds of others, and tragically there will be hundreds more in the future.

There is a popular misconception that the ecology movement is a peripheral cause unconnected to the important issues of national politics. Until recently, it has been thought of as 'something to do with pandas and muesli'. Yet, the loss in terms of human lives alone is appalling. Man's exploitation and pollution of an environment in which even the most basic ecological concerns – such as clean air and clean water – are ignored has resulted in human casualties on a massive scale.

In order to grasp the extent of this tragedy, and the range of the conflict, let us build a monument in our imaginations to the cost in human lives alone.

I suggest we construct a pyramid from the coffins of those who are killed in the ecology wars in just one year. For the sake of illustration, we will build it in reverse order, in six stages from pinnacle to foundation.

1 At the peak of the pyramid we place the coffins of the most conspicuous casualties: the scores of conservationists who are murdered outright for their stand against the destruction of the environment and other species.

2 Next we have the coffins of the hundreds of tribal people who are massacred because they occupy and protect wilderness lands which others wish to exploit.

3 Beneath these are the thousands of coffins of those drowned in the many floods which directly result from ruthless cutting-down of mountain forests.

4 Then we come to the tens of thousands of coffins of those who die through chemical poisoning, toxic waste pollution, atomic radiation, and industrial fires and explosions.

5 Further below are the coffins of those who die in droughts and famines brought on by farming and grazing methods which result in soil erosion and desertification.

6 Finally we come to the massive base of the pyramid. It is built of over twenty-five million coffins for those people who are killed through drinking and using polluted water.

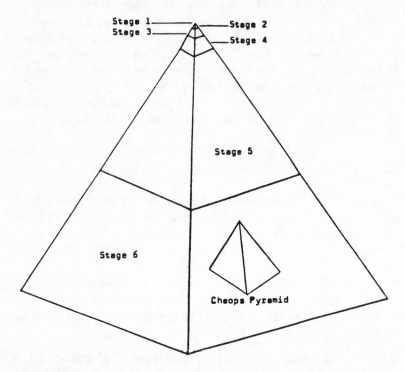

The world's largest pyramid was by King Cheops in ancient Egypt. It took a hundred thousand slaves over twenty years to build. The monument we have just constructed, representing the fatalities of the eco wars *for only one year*, is some ten times the size of the Cheops pyramid.

Although the eco wars result in a death rate greater than one Hiroshima a day every day of the year, extraordinarily, most of the squabbling nations of the world place ecology very low on their list of priorities.

But the eco wars are concerned with more than human losses. The primary casualties are the animal and plant species that are threatened with extinction. If these were also taken into account, our colossal human pyramid would in fact be only the tiny tip of another staggeringly huge pyramid representing the total biomass destroyed each year.

However, statistics on this scale are overwhelming; they tend to paralyze people rather than activate them. We cannot absorb the impact of millions of deaths, and I have only made use of statistics here to demonstrate a point about human priorities. For the most part, this book is filled with extraordinary stories: some appalling, some heartening, some funny. It is a book about some small miracles, and even a few large ones. It is a book about talking apes and singing whales, assassins and pirates, cactus-rustlers and orchid smugglers, animal martyrs and green-thumbed saints, paratrooper elephants and tracker cockroaches.

It is concerned with individual actions and individual deeds, and it aims to show such actions and deeds do count. In fact, it proposes that this is the only way in which change can come about: through one person acting and changing one small part of his world.

This book is about people who have taken to heart Edmund Burke's dictum: 'Nobody made a greater mistake than he who did nothing because he could do only a little.' The guerilla soldiers of the eco wars are engaged in issues that are sometimes small, sometimes large; but whether it is the destruction of a rare egg or the destruction of the planet, the principle behind the issue is the same.

As in any guerilla war, there are numerous factions, each with its own priorities and methods. The movement, although predominantly non-violent, encompasses every position from terrorism to pacifism; every tactic from bombs to legislation.

The Environmental Wars is a report from the war zones of a global conflict that in one way or another involves each of us, every day of our lives.

1

Assassins and Martyrs

Notable conservationist murder cases

'For as long as men massacre animals, they will kill each other.' *Pythagoras*

1

Tarzan's Best Girl

Dian Fossey • Killings and mutilations of mountain gorillas of Rwanda • Tourist market for hands and skulls • Fossey's attempts to control killings • Conflicts and threats • Fossey's murder • Gorilla graveyard.

On New Year's Day 1978, a researcher on patrol in Rwanda's Park of the Volcanoes found the mutilated body of a male gorilla in the forest undergrowth. It was a gruesome sight. The body had multiple stab wounds from spear thrusts, and had been grossly mutilated. Like the victim of some bizarre torso murderer, the young silverback gorilla's head and both its hands had been savagely hacked off and carried away.

The event made headlines all over the world, for this was not just another anonymous animal which had died violently in the central African wilderness but an animal celebrity, known to millions of people all over the world.

The gorilla was called Digit, a name given him by the gifted and dedicated American scientist, Dian Fossey, who over the previous eleven years had studied and befriended his family group in the Rwandan forest. Fossey had known Digit since birth, ten years before. He had become a favourite of hers and featured in many television films and *National Geographic Magazine* articles about Fossey's work with the mountain gorillas. A large poster with a photograph of Digit had been circulated worldwide by the Rwandan tourist bureau. The caption on the poster read: 'Come and see me in Rwanda.' He was, without question, Rwanda's most famous son.

A few days later one of the half-dozen men who had taken part in this barbaric act was captured by Dian Fossey and her workers. Eventually, the poacher was turned over to the police, but not before Fossey had extracted an account of the killing.

The gorillas had been pursued by the hunters with dogs for two days. Exhausted at last, they were brought to bay. As one of the dominant males, Digit turned on the hunters and fought a rearguard action in order to allow the safe retreat of the others. It had taken five spears to bring the young gorilla down and, even so, he had killed one of the hunting dogs before he died. But why, Fossey wanted to know, had the animal been killed, and why had it been mutilated?

The hunter's answer stunned Fossey. They had been offered ten pounds by a trader for the head and hands of a silverback gorilla. To what purpose? The gorilla's hands would be made into ashtrays and the skull into a trophy for sale to tourists. There were only two hundred and fifty mountain gorillas in existence in 1978 and yet one of them had been slaughtered for the sake of ten pounds and three pieces of tasteless tourist junk. It was just another action in a long series of events which brought the species one step closer to extinction.

In 1960, the mountain gorilla's plight was already precarious, with an estimated population of five hundred in a park measuring some twenty-five miles by six. However, it became dramatically worse later in the decade when the World Bank financed a project which effectively destroyed 40% of the park forest in order to cultivate pyrethrum, a daisy-like plant used to manufacture an insecticide which was marketed in Europe. This habitat destruction brought the gorilla population down to about three hundred.

However, with the diligent work of Dian Fossey and a few others who patrolled the forest in order to protect the park habitat from poachers and encroaching cattle, the gorilla population was thought to be stabilizing itself to a degree. Despite setbacks, it seemed that Fossey's combination of 'active conservation', as she called it, and international publicity, might very well save the species from further decline.

Before the macabre market in gorilla trophies opened up, gorillas were mostly killed by cattle herders or strangled by wire nooses set to snare other animals. Now, however, Fossey estimated that nearly twenty gorillas a year were being killed to satisfy this demand for trophies from white tourists and collectors. In one year alone, thirty skulls were seized by officials. Furthermore, unscrupulous European zoos with the help of local officials were still financing the illegal capture of live infant gorillas, despite the fact that such captures could only be made by killing the mother and at least one dominant male.

Digit had been killed less than three miles from Dian Fossey's base camp. Seven months later, a little further up the slopes of Mount Visoke, two more gorillas from Digit's family group were found similarly killed and mutilated. These were the female called Macho and the big male silverback, known as Uncle Bert. This time high-powered rifles had been used.

Poachers had already slaughtered large numbers of the elephant, buffalo and various monkey species in the park region, so, with the killing of Digit and others of his tribe, Dian Fossey adopted an altogether more radical stance. She believed that the time for passive observation of the gorillas had long gone, and from now on she became more and more involved in the protection of the species.

Dian Fossey's drive to save the mountain gorillas was formidable. The publicity of the gorilla killings had resulted in wide sympathy for Fossey's idea of active conservation and she established the Digit Fund in America to finance regular full-time patrols in the park. The fund helped to some degree, but she often had to fall back on her own meagre financial resources to keep the patrols going. These patrols succeeded in collecting thousands of traps and snares which had been illegally set to capture the animals. Furthermore, the patrols actively drove off, and even captured, many poachers within the park. There was no doubt whatever that they were successful in keeping the poachers at bay.

Perhaps Dian Fossey's campaign was just too successful, as Farley Mowat suggested in *Virunga*, his biography of her. During the last year of her life she uncovered a network of sixty poachers and a score of Pakistani and white dealers. She also stumbled upon some sort of gold and ivory smuggling operation. Mowat quotes Fossey's journal: 'There are names on the list that could mean big trouble if released.' Big trouble indeed. A curse of death by black magic, in the form of a puff-adder fetish, was planted in front of Fossey's cabin. Shortly after, the captured poacher who had informed Fossey about the network was nearly killed while in prison. These and other sinister incidents lead one to conclude Fossey was treading on dangerous ground.

At the same time the actions of what was essentially a private conservation army, led by Dian Fossey, were antagonizing certain members of the Rwandan government and others within the conservationist community. After all, they argued, there were already park guards and government officials to protect the gorillas. Fossey's frank reply was that they were extremely ineffective. She believed there was considerable corruption among officials and that the poorly paid guards were an easy target for bribery.

One source of conflict which was especially upsetting to Dian Fossey came, somewhat unexpectedly, from the Mountain Gorilla Fund. This fund had initially been set up by the Fauna Preservation Society in Britain in the wake of publicity generated by Digit's killing, and it was the belief of Fossey that the money would be used to subsidize her activities. However, she was given no control of the finances of the fund at all, and the end result was the setting up of a rival organization, the Mountain Gorilla Project (MGP), whose administrators were extremely hostile to her.

Another source of conflict was the Office of Rwandan Tourism and Parks (ORTPN) which, before Fossey's arrival, was only dimly aware of the gorillas in its forests. By the time ORTPN realized that they were sitting on a tourist gold-mine they had the fiercely independent Fossey to contend with. Her single-minded determination to protect the gorillas often embarrassed ORTPN officials and they preferred to deal with the more flexible MGP. Before long plans were afoot to put all management of the gorillas into the hands of the Mountain Gorilla Project. Fossey's research station was to be turned into a tourist facility, and Fossey herself was to be pushed out of the park and the country.

Despite the odds against her, including her own precarious health, Dian Fossey persisted in her battle to protect the mountain gorillas. By now she had become a fervent and uncompromising fighter for the gorillas' right to life, and once that choice was made, she pursued it to the end. Her state of mind was perhaps best conveyed by the statement Farley Mowat discovered on the last page of her journal: 'When you realize the value of all life, you dwell less on what is past and concentrate more on the preservation of the future.'

On the night of 26 December 1985, a terrible struggle took place in Dian Fossey's remote cabin at Karisoke research centre. It is uncertain whether her attacker was let into the cabin or forced his way through the corrugated metal sheathing at the back. What *was* clear was that Fossey had put up a very brave fight: at the time of her death she was frantically trying to load an automatic pistol with a clip of the wrong calibre bullets. At that point, it seems, her attacker struck her repeatedly with a large and heavy native knife, called a *panga*.

The savagery of the attack was shocking. There were several wounds to the head, but the fatal blow literally split her skull open. It was a horrendous eight-inch gash which ran from the forehead across the nose and down the cheek to the edge of her mouth.

Controversy continues as to the identity of Dian Fossey's assassin. And assassin he certainly was, for although the cabin had been ransacked to make it look like an attempted robbery, nothing was stolen, despite the fact that there were several thousand dollars in cash, traveller's cheques and camera equipment for the taking. There have been many theories and speculations about Fossey's murder. In the end, the Rwandan authorities chose a scapegoat in the form of one of Fossey's American research assistants, Wayne McGuire, who was charged and convicted of the crime in absentia. McGuire vehemently denies the charge, and very few people find the accusation believable.

Regardless of who committed the act, there is no mystery at all as to the reason for Dian Fossey's murder. In a graveyard she herself had cleared near her cabin at the park's research station are buried Digit, Macho, Uncle Bert and eleven other apes who were slaughtered in this little African bush war against the mountain gorillas. Dian Fossey is now also buried in that graveyard; a soldier whose life was lost in the front lines of that war.

At her graveside, the missionary Elton Wallace was inspired to deliver a moving eulogy. Noting that Fossey's death had taken place the day after Christmas, he felt compelled to draw a parallel: 'We see at our feet here a parable of that magnificent condescension – Dian Fossey, born to a home of comfort and privilege that she left by her own choice to live among a race faced with extinction . . . she will lie now among those with whom she lived, and among whom she died. And if you think that the distance which Christ travelled to take on the likeness of Man is not so great as that from man to gorilla, then you don't know men. Or gorillas. Or God.'

2

The Cat Woman

Joy Adamson • Primatologists Jane Goodall, Dian Fossey, Birute Galdikas: the Trimates • Kidnapping of Goodall's researchers • Adamson's book, *Born Free* • Leopard killings and other scandals • Kenya's Shaba Lion preserve • Adamson's murder.

When Dian Fossey first went out to Africa in 1966 to study and live among the mountain gorillas, she was following the example of another brave woman, Jane Goodall, who by 1966 had already been studying chimpanzees in the Gombe Preserve in Tanzania for six years. The British Goodall, the American Fossey, and the Canadian Birute Galdikas, who studied orangutans, were three women primatologists who observed the behaviour of the great apes by coming to be accepted, more or less, as part of the animals' tribes. Theirs were the first really long-term studies of primates in the wild, and had all been engineered by the famous paleoanthropologist Louis Leakey who was convinced that primate research was essential to an understanding of early human evolution.

As Leakey had intended, the work of all three women – with its Tarzan-like concept of becoming a part of the ape tribe and striking portrayal of beauty-and-the-beast morality – captured the public imagination. These three primatologists, or the 'Trimates' as they were sometimes jokingly called, developed a greater public awareness of, and interest in, the great apes than anyone had before them. These attractive women – the blonde Goodall, the dark-haired Fossey, and the auburn Galdikas – and the undeniably gentle and engaging objects of their studies, became a part of popular folklore through the modern media.

Fortunately, Goodall and Galdikas have not become victims of violence like Dian Fossey, although both have met with considerable obstacles. Jane Goodall, like her friend Fossey, has suffered from intimidation and threats to the continuation of her research: the worst incident occurring in May 1975 when three of her students from Stanford University and one Dutch research assistant were kidnapped from the Gombe Stream Preserve and taken across Lake Tanganyika to an armed camp in eastern Zaïre. After five months of negotiations, a ransom was paid and the hostages were released.

Another brave and independent woman in Africa whose work embodied a variation of the beauty-among-the-beasts theme was Joy Adamson. Adamson's books, including *Born Free*, were worldwide bestsellers and were made into two motion pictures. Adamson's chosen beasts were the big cats of Africa: the lions, cheetahs and leopards, and her accounts of her life and times in the African wilderness did much to reverse the image promoted by the big game hunters of the big cats as cruel and merciless killers. In fact, her books succeeded in furthering the cause of conservation for all species.

However, like Fossey and most others who are exposed to the atrocities of the traffickers, Adamson's attitudes to those who were devastating the wildlife populations hardened into angry opposition. Many of her public exposures of illegal trade in wild animals implied extensive government corruption. Some very obvious conclusions, for instance, might be drawn from the fact that in one year in East Africa only five hundred legal permits for killing leopards were issued, and yet customs officials allowed over fifty thousand leopardskins to be exported to Europe.

Furthermore, Joy Adamson did not endear herself to the traffickers or to the fur industry by disclosing the horrific way in which many of the big cats were being killed. To ensure that leopardskins were not damaged by bullets, for instance, poachers were trapping the animals alive in small cages. To minimize the pelt damage, the caged animals were then killed with a red-hot iron poker or spear driven up the anus.

Increasingly, Joy Adamson worked to preserve at least a small part of the wilderness she had come to love, and she found this more and more difficult to achieve.

On 4 January 1980, Joy Adamson was found dead on the Shaba Lion Preserve just north of Nairobi in Kenya. Initial reports suggested that she had been mauled to death by lions. However, further investigation revealed that she had been murdered by a Turkana tribesman. She had been brutally stabbed and badly torn up with a *simi* – a short curved knife – in an attempt to make it appear as though she had been killed by those same animals she had worked so hard to save.

3

The Bird Man

Guy Bradley • Egret slaughter in Florida • Shooting of Bradley • Extinction of bird species for feather fads • Legislative action • New York's millinery market for feathers and its collapse.

Dian Fossey and Joy Adamson are just two of the more well-known casualties in a secret war that has been gaining momentum ever since 8 July 1905.

On that fateful day, an Audubon Society warden in the Florida Everglades took his skiff out to a schooner anchored near Oyster Key, off the tip of the Florida mainland. Thirty-five-year-old Guy Bradley was one of four wardens employed by the newly-formed society in the Everglades to enforce the laws banning the slaughter of critically endangered egrets.

When Bradley reached the schooner, he discovered two hunters loading dead egrets aboard. When he attempted to arrest the hunters, the ship's skipper pulled out a gun and shot the warden at point-blank range. The schooner fled, leaving Bradley dead and adrift in his skiff. His body was discovered by two boys the following day. Bradley's murderer, through some extraordinary legal manoeuvring, was allowed a claim of self-defence and was subsequently set free.

However, the tragedy of Bradley's murder provided considerable momentum in the drive to have egret feathers banned in New York's millinery trade. The incident helped to turn the public against the wanton slaughter of these beautiful birds for the sole purpose of supplying feathers for ladies' hats.

At the time of Bradley's death, the millinery trade was responsible for the killing of five million birds annually. Its demand for exotic plumage to decorate hats created a market for every kind of feather, from the tiny hummingbird to the huge whooping crane. A sudden craze for a particular type of exotic feather could result in a killing frenzy of remarkable ferocity. A case in point was the unique New Zealand bird called the huia, one of the country's most remarkable species. Always a rare bird, the nineteen-inch-long huia was prized by Maori chieftains for its long black tail feathers with their distinctive white tips. The feathers were worn only by high-ranking chieftains as a symbol of authority. Hunting was carefully limited by tribal laws.

Unfortunately, at the turn of the century, the Duke of York – the future King George V – made a royal visit to New Zealand where a Maori chieftain presented him with the elegant black-and-white feathers of a huia. The Duke promptly stuck the feathers in his hat and, by that single action, promptly condemned the huia to extinction. An immediate fashion for huia feathers followed the Duke's return to England. By 1907, the last huia had been shot and plucked.

Although the egret was far more common than the huia, egret feathers were in much greater demand, and for much longer, than those of the huia. In one season, in Florida alone, 130,000 egrets were slaughtered on their nesting sites. Egret plumage, with long, silky filigrees of trailing feathers, was the most sought after of all millinery decoration. Indeed egret feathers were sold by the ounce, at quite literally the price of gold. At the turn of the century, they were selling for the then-phenomenal price of $38 an ounce.

Ironically, the shot that killed Bradley also delivered a death blow to the trade in egret feathers, for it prompted the president of the Audubon Society, William Dutcher, to point the accusing finger. 'Heretofore, the price has been the life of birds. Now human blood has been added.' Soon after Bradley's death, laws were pushed through in New York, the centre of the millinery trade, prohibiting the use of wild bird feathers in the decoration of hats. Thus the egret was saved from extinction largely through the martyrdom of Guy Bradley.

4

The Elephant Man

Roger Edward • Africa: gunfight in the elephant
reserve – seven die • Other gunbattles in Tanzania,
Zaïre, Uganda, Kenya, Angola • Zaire's valley of
volcanoes – 23 wardens killed • Government and
army involvement.

Since Guy Bradley's murder in 1905, the protection extended towards
many endangered species has improved dramatically. However, the
fact remains that even in the best-policed nations in the world today,
illegal hunting of protected and endangered species in nature reserves
and national parks is still a multimillion-dollar black-market industry.
In France, for instance, in the last five years, two federal guards were
shot dead and two others wounded by poachers while attempting to
protect the last thirty Pyrenean bears in the world in their only
remaining mountain reserve.

In Africa, particularly when it comes to elephant ivory, we are not
dealing with small-scale skirmishes, but a highly organized, continent-
wide guerilla war involving thousands of mercenary hunters and
traffickers. Consequently the protection of the elephant herds of Africa
has resulted in the highest murder rate of all among conservationists.

On the Zimbabwe national elephant reserve, in May 1980, chief
game warden Roger Edward approached a group of men trekking
through the parkland. When Edward came within range, the men
suddenly produced AK47 automatic rifles and instantly cut down and
killed Edward and another warden. Other wardens returned the fire,
and an ensuing running gun battle between wardens and poachers
resulted in a scene reminiscent of a Wild West shootout. The bloody
confrontation ended with the violent deaths of seven men.

Edward was just one of the scores of conservationists who have died
in the elephant war. In Zambia, over a seven-year period in one reserve
alone, ten game wardens were murdered. A year before Edward's
killing, in June 1979, poachers opened fire with sub-machine guns in
the Naorongoro conservation area in Tanzania and instantly killed the
country's chief warden.

In Zaïre, in 1975, there was a celebration to mark the fiftieth anniversary of Africa's first national park in the Valley of the Volcanoes – a wilderness that extends over the border to the Rwandan park where Dian Fossey's gorillas roamed. No fewer than twenty-three game wardens had been murdered in the fight to protect the park's wildlife during the half-century and, as a tribute to them, a bronze plaque engraved with their names was put up by the World Wildlife Fund.

It is not just bandit armies that loot Africa's heritage for their own profit; corrupt government officials, and many at the head of the military forces, have become rich through the slaughter of elephants and other wildlife. In 1980, when the government of the Central African Republic's self-proclaimed Emperor Jean Bedel Bokassa collapsed, under the weight of its own outrageous extravagance, one source of Bokassa's wealth emerged. Bokassa's army had been ordered to take out helicopter gunships and machine-gun no less than thirty thousand elephants for their ivory. In Uganda, President Amin's army was similarly employed in what amounted to the virtual extinction of the elephant in that country. Both countries faced human starvation on a massive scale, yet no attempt whatever was made to use the meat of the thousands of slaughtered elephants to alleviate this. In both places, the mass killings simply made a few men rich enough to buy weapons which were used to kill and intimidate their own people.

In Uganda, the few park rangers who survived the purges of the Amin regime, and the subsequent raids by Tanzanian soldiers, were literally reduced to wearing rags and carrying worn-out bolt-action rifles. On one occasion, two surviving rangers on an elephant and rhinoceros reserve had only one operable bolt-action rifle between them. With only fifteen rounds of ammunition left, their job was to defend the animals from a known local army of at least four hundred poachers armed with automatic weapons and equipped with modern jeeps and a helicopter.

Whenever political instability undermines the capacity of an African nation to protect its parks, there are those who make their fortune by engineering the looting of the reserves. The South Africans are particularly adept at this kind of vulture-like behaviour. In Angola, the South African-backed UNITA armies run a profitable sideline in their war, often using military helicopters and vehicles to smuggle Angolan ivory into neighbouring South Africa. Similar poaching and smuggling schemes using South African military personnel and aircraft have been in operation in Namibia, Mozambique, Botswana, and Zaïre. It goes without saying that anyone interfering with the military precision of this large-scale black-market industry is quickly eliminated.

5

The Soviet Soldier

Valéry Rinchinov • Night hunters at Lake Baikal •
Killing of Rinchinov • 'Official' poaching and
government leniency • Caviar-smuggling case •
Marshall Chuikov's hunting party • Marshall
Batitskii's polar bear hunt • Soviet black market in
meat and furs • Military involvement.

It is not just the nations which advocate rampant free-enterprise systems that are open to the corruption of illegal trade in endangered and protected animals.

In March 1982, forest ranger Valéry Rinchinov, acting on his own initiative, went into the Siberian forest nature reserve near Lake Baikal to check out reports of illegal night hunting. With two friends and a driver, the 24-year-old former marine saw hunters' lights on the far side of a clearing. As he walked towards the lights, a spotlight suddenly shifted and shone on him.

As one of the ranger's friends, Comrade Amogolonov, later recounted: 'A shot rang out as we heard the cry of the wounded Rinchinov.' The poachers left Rinchinov where he lay in the snow and fled by car. By the time Rinchinov's companions reached him, he was already dead. However, Amogolonov was just able to read the licence number of the poachers' car before it disappeared.

This action soon led to the discovery of the killers. However, it was a highly embarrassing situation for local officials: one of the poachers proved to be the first secretary of the regional Young Communist League. Another was the director of the regional film distribution network and the third was a government wholesale official. The fourth man was no less a personage than an officer of the district hunting industry.

Outrageous though this murder-scandal was, perhaps the most unusual aspect of the Rinchinov affair was that it was reported in Pravda, the USSR's official communist daily newspaper, in a tone which conveyed indignation at the workings of Soviet justice. The report pointed out that, apart from the removal of two of the officials from their posts, the killers went unpunished for their crimes.

Until President Gorbachev's reforms a few years later, such admissions of Soviet injustice were rare. It was also rare that the Soviet system was so lenient on its transgressors. In April 1982, for instance, Soviet police uncovered a multimillion-dollar caviar-smuggling operation. Over two hundred employees in the fisheries and foreign trade ministries were charged with fraud and given severe sentences. Nor did senior officials escape prosecution. The Deputy Minister of Fisheries, Vladimir T. Rytov, was also convicted. He was sentenced to 'an unusual measure of punishment' by the courts, the current Soviet euphemism for execution by firing squad.

In the caviar-smuggling case, a good deal of money was involved, but at least no human blood had been spilled. So why, in the case of the murder of Valéry Rinchinov, was there such leniency? The answer was simply that there was far too much to cover up. If the minor party official had received the punishment he deserved, they might have attempted to point out how common the practice of poaching was with all party officers. They would doubtless also have claimed that it was not they who were behaving unusually, but the impulsive young ranger, Rinchinov, acting 'on his own initiative'. Had he checked with his superiors, he would certainly not have been permitted to go looking for illegal night hunters.

The simple truth is that nature preserves and sanctuaries in the Soviet Union are used as communist party-members' hunting grounds, and as a source of black-market meat and furs in a widespread network of corruption. As the Soviet dissident scientist, Boris Komarov (pseudonym), reported in *Destruction of Nature in the Soviet Union*, published in 1977: 'A ban on hunting has been violated by the very pillars of society, from the district level on up. This partly explains the lenient attitude of the authorities toward poachers.'

Indeed, high-ranking government and military commanders were legendary in Russia for their hunts on nature preserves. In one famous hunt, the then-chairman of the Council of Ministers for Civil Defence, Marshall Chuikov, went with friends and military cronies to a nature conservation park in the Kyzl-Agach preserve. In characteristic Party style, the entourage drove in a column of military all-terrain vehicles, followed by communication cars and field kitchens.

As the vehicles flew over the banks and shallows, ducks, geese and teal were shot to pieces by the concentration of heavy gunfire in the shallow inlets. After a break for luncheon, Chuikov and his comrades transferred to helicopters, from where the main prey was soon spotted: herds of wild boar. The helicopters swooped down and shot the herds out of existence with heavy-gauge automatic rifle fire. In the end, even tanks were brought into the preserve. In their enthusiasm the hunters had accidentally bogged down the all-terrain vehicles in the swamps. The tanks were needed to drag them free.

For the most part, the excesses of the military in the USSR are scrupulously ignored. However, in 1976, a series of embarrassments concerning the black-market trade in the furs and skins of endangered species forced the Party to act against one of its most prominent military men, Marshall Batitskii, Commander of the Strategic Missile Forces in the Arctic Archangel Sector. Marshall Batitskii was known to have hunted and shot at least fifteen protected and endangered polar bears and in a rather flamboyant manner. He casually borrowed a military helicopter gunship and used its specially constructed machine-gun turret to bring the bears down. Furthermore, by covering the floors and walls of his home and office conspicuously with the skins, Batitskii made it virtually impossible for other military and government officials to ignore his 'hobby'.

However outrageous the reported incidents have been, the scandals exposed are obviously only the tip of the iceberg. Four years before the murder of Rinchinov, in 1978, a report appeared in the Soviet *Literaturnaya Gazeta* claiming that there had been scores of deaths in pitched battles between Siberian game wardens and Soviet soldiers using helicopters, automatic weapons and hand grenades to poach deer, wild boar, tiger, bear, and wild sheep. Boris Komarov confirms this in his study, pointing out that the army, which is best equipped for such activities, finds it easy to pursue them for unspecified reasons of 'strategic necessity': 'More than once game wardens in the Chita and Amur regions have stopped army trucks loaded with dozens of carcasses of stags, deer and elk. But the wardens were powerless if the poachers had time to stick a 'radiation' label or a 'secret cargo' sign on the truck body.'

6

The Green Rangers

Fighting back • Armed conservationist enforcers:
Green Men of Kenya • Warden Ted Goss •
Exchanging fire • A crucifixion • Front-line troops.

Those who are murdered in these battles to save endangered species occupy only one of the many front lines of the ecology wars. There have been martyrs in all of them, but there seems to be a kind of selfless purity in those who risk their lives for a cause of no direct benefit to themselves or even to their own species. They hold the belief that the extinction of species at the hand of man is an ultimate crime against the whole of life upon the planet.

Increasingly, since the 1960s popular feeling has been slowly changing the attitude of governments towards nature and its creatures. Many nations have found it necessary to recruit and arm soldiers to protect endangered species from further slaughter. In the mid-1970s, the Kenyan government recruited a small army of bush-rangers to combat bandit gangs of organized poachers on government reserves. Because of their green camouflage jackets, the rangers soon became known as the 'Green Men'. This force was led by game warden Ted Goss, who claimed some considerable success at cutting down the rate of poaching despite the barriers created by high-level government corruption.

In one of many cases, the director of Kenya's national parks was removed from his post after he was found to be involved in a smuggling racket along with many members of his staff. Further prosecution was not deemed wise because too many people within the government and within high-placed Kenyan families were involved with the smuggling of ivory.

Ted Goss cheerfully acknowledged the highly organized and well-armed nature of the poacher armies. He claimed there are certain advantages in this state of affairs. 'It is much easier to find the poachers nowadays because they shoot at you. They blaze away at my helicopter whenever they see it.' His devil-may-care humour, however, rather belied the serious nature of his work.

Since 1980 the conflict has become ever more deadly. The human and animal casualties have rapidly escalated, and groups like the Green Men are fighting rear-guard actions. In Kenya – and throughout Africa – the anti-poaching patrols and the National Parks are critically underfunded, and the outlaw armies go where they please.

In the first year of operation, two Green Men were murdered, and others soon followed. This is not to say that the Green Men were taking their punishment lying down. Far from it.

'They say we are rough on poachers. This may be so,' admitted Goss, who often acts as a restraining influence in this conflict. After a recent murder of a ranger, it seemed that some of the Green Men took a rather vengeful view. In the autumn of 1980, a Nairobi newspaper reported that a number of Green Men were suspected of inflicting very rough treatment indeed on a poacher. The newspaper reported that the man had been found alive, but nailed to a tree 'in just the manner Jesus Christ was crucified'.

Making no excuse for atrocity on either side, the Green Ranger Ted Goss would just nod sagely, as he often did, and say, 'Make no mistake. We are fighting a war.'

2

Traffickers and Mercenaries

Illegal trade in endangered species

'There are some occupations that make it impossible for a man to be virtuous.' *Socrates*

1

The Great Egg Robbery

One man robs British Museum of thirty thousand
rare eggs • Black market in eggs and birds • Other
egg thefts and smuggling operations • Plight of the
condor and other endangered species.

The British Museum's ornithological division at Tring in Hertfordshire
boasts the finest collection of rare birds and eggs in the world. It is a
massive reference library for natural historians and artists who come
to research into many rare and even extinct species.

In 1978, one regular amateur researcher was taken to one side by the
security guards at the exit door of the museum. A search of the suspect
soon revealed that beneath his loose-fitting trousers he wore a
women's panty-hose stuffed with nearly two dozen rare eggs. The man
had been visiting the museum several times a week for more than two
years and, as further investigation revealed, he had been steadily
looting the museum during this period.

The scale of the theft stunned the museum staff. It was the greatest
robbery of rare eggs in history. When an audit of the massive
collection was made, it was discovered that the man had stolen no less
than thirty thousand eggs! Furthermore, hardly any of these eggs
could be recovered. The man was not just an eccentric magpie stealing
eggs for his own collection; he was a full-time professional egg robber
who was a major trader in the egg-collectors' black market.

The robbery was a disaster for the museum. Many of the eggs were
irreplaceable as they came from extremely rare or even extinct birds.
On top of this, the museum was faced with the enormous and costly
task of identifying and re-cataloguing what the thief had left behind,
for to cover his tracks, he had rearranged eggs and switched labels on
many specimens and boxes. The result was the 'contamination' of
every section he had looted. It would take years of work just to get the
depleted collection back into some sort of usable order.

This bizarre robbery from the British Museum demonstrates the voracious appetites of the wildlife market, even in its more obscure areas. But whereas the eggs stolen from the British Museum were already 'dead', most of the millions of eggs stolen each year are taken 'live' from the wild. In the case of plentiful species this is not a major concern, but inevitably the rarer the species, the more it is worth and the more fiercely it is hunted. It is a strategy with an inevitable and deadly result. As with the plumage of birds, when supply diminishes, demand increases.

A variation on this egg-collectors' market involves various species of rare falcon which fetch huge prices in Arab nations. The rewards are so large that they are difficult for some of the more creative smugglers to resist. One such individual, after establishing the fact that a baby gyrfalcon could be sold in Saudi Arabia for $80,000, was arrested while attempting to smuggle a dozen gyrfalcon eggs out of Canada in a suitcase fitted as an incubator.

In cases such as the California condor, which survives with a population of exactly twenty-seven, collectors can still be found who will pay $2,000 and more for a single stolen egg. Unlike the falcon smugglers, these people are not interested in hatching the eggs but only in preserving the shells. For the sake of their collections, they are willing to put one more nail in the coffin of the species. From the mid-nineteenth century onward the collectors of rare eggs and skins for ornithologists have been responsible for numerous extinctions. When they see an endangered species struggling to cling precariously to life, they have no sense of the tragedy of such a loss. On the contrary, there is a delight in hunting it down, for the bird's extinction will make the egg more sought after than ever. It will dramatically increase the value of their investment.

A recent Sotheby's auction of a stuffed Great Auk to the Icelandic government gives a weird perspective on the strange values we place on wildlife. Egg-and-feather merchants from Iceland were largely responsible for the extinction of the large penguin-like bird in the North Atlantic. Now, long after its demise, the same government through popular subscription paid the highest price ever for a stuffed animal just to have one example of a species that it once possessed by the million. Perhaps the final irony was the fact that the price paid for the stuffed bird was greater than the total profits of the egg-collecting industry that exterminated the species.

2

The Skin Trade

Billion-dollar black market in animal skins • New
York fur-smuggling ring bust of 1973 • West
Germany, US and Japan; top consumers • Prime
victims: spotted and striped cats • Eco-cops and
robbers: 1973 formation of CITES (Convention on
International Trade in Endangered Species)

One of the most lucrative areas of wildlife commerce is the fur trade.
Over fifty million fur-bearing wild animals are killed each year to
supply the industry. This is a volatile issue requiring immediate
attention and legislation, particularly in relation to the trade in the
skins of the wild cat family. Virtually all of the big cats should have
protected status.

In 1973, the United States Customs exposed a massive smuggling
operation involving a ring of thirty-three furriers who were caught
smuggling a shipment of 250,000 skins of animals on the endangered
list. A New York fur dealer, Vesely-Forte, was found in one year alone
to have handled over 100,000 illegal pelts worth a total of five million
dollars. These included 30,000 ocelots, 46,000 margays, 6,000
leopards, 2,000 cheetahs, 2,000 jaguars and 15,000 otters. This
prosecution came as a considerable shock to many who believed
America had the best enforcement laws against the trade in
endangered species. Even more shatteringly the operation merely
revealed the tip of the iceberg.

West Germany is the world's largest buyer of furs, with the United
States and Japan coming second and third respectively. In these
countries the fur trade is a powerful billion-dollar-industry determined
to protect itself against control by government or public opinion. Its
sub-industry, the black market in illegal skins, is even more ruthless
about protecting its business interests. And the fact that the 'legitimate'
trade and the black marketeers work hand in glove is highly
advantageous to both.

In Japan, for instance, despite the banning of fur sales for many
years, department stores were still openly selling full-length tiger and

clouded-leopard coats well into the 1980s at prices in excess of $100,000 each. Even today, advertisements for the sale of coats made from the incredibly rare snow leopard can still be found in Tokyo newspapers.

In West Germany, in 1977, one dealer alone imported 75,000 endangered wildcat skins from Brazil under false licensing. The same year another dealer, also from West Germany, illegally imported from Paraguay a further 400,000 endangered-status skins worth more than $12 million.

But again, we are dealing with just the visible manifestations of this trade. The disparity between official and actual figures is evident to anyone even briefly observing import and export figures. Typically, in one year when the official world export figure for wildcat skins was 60,000, customs records show that over 300,000 wildcat skins were imported into West Germany alone during that period.

The impossible complexity of the international trade in animal species led to the formation in Washington, in 1973, of a treaty organization with a mandate to organize and control the movement of animals and animal products to world markets. This organization is known as the Convention on International Trade in Endangered Species – CITES. It currently has approximately a hundred signatory members – nations, which, in theory at least, have agreed to prohibit international commercial trade in over six hundred endangered species.

In a few areas, CITES has had some effectiveness in standardizing regulations concerning the animal trade in member countries. However, since penalties for the violation of CITES regulations are left to the member nations, the influence and lobbying power of this billion-dollar-industry has resulted in ludicrously light penalties for violators in most countries. Added to this is the problem created by non-member nations, where anything goes.

The difficulty with CITES is that its very existence tends to lull governments and the general public into assuming that the traffic in endangered species is being properly controlled and policed. This has not even begun to be true. For the most part the animal trade is uncontrolled and policing non-existent. Forged CITES certificates are easily bought on the black market, and most customs authorities have no training whatever in species identification.

After fourteen years in operation, at the 1987 convention, CITES officials admitted that their organization was only about 17% effective in controlling trade in endangered species! This is partly because of its own priorities: it is after all, primarily a trade convention, not a conservationist organization, and it suffers from the problem of legitimizing some very dubious traders. It also legitimizes markets in some species which end up being used to provide routes for laundering prohibited species.

CITES is a necessary beginning in the control of the trade in endangered species, but it has many obstacles to overcome before it can even hope to control and police an expanding industry. In the endangered species skin trade, as in many other areas in which CITES deals, crime still pays very well indeed.

3

The Slave Trade

Smuggling of live animals, one hundred million dollars a year • Jean-Yves Domalain: confessions of a trafficker in SE Asia • Systematic bribery, deception, fraud • Cruel and wasteful methods of trapping primates and other species • Appalling conditions of captivity and shipping • Pet, circus and zoo trade • Zoo as smugglers' front.

Often working hand in hand with the black marketeers in the fur and skin trade are those who deal in the illegal and hugely profitable business of selling live endangered species. In the United States alone, the black market in such animals amounts to a hundred-million-dollar-a-year business. This is generally a fad market for exotic pets purchased by people with no idea of how to care for them properly. More than 90% of the animals purchased are dead within six months of captivity.

The recent American fad for exotic birds, particularly macaws from South America and cockatoos from Asia, at one point resulted in retail prices that exceeded $10,000 for each bird. Another American fad, for unusual reptiles, resulted in a demand for one of the world's rarest, the tuatara. The problem was that in New Zealand, where the only tuatara population survived, this unique and critically endangered creature which evolutionarily predates the dinosaurs was considered a national treasure protected by well-enforced laws. However, such is the lure of the marketplace that a price-tag of $10,000 or more for each reptile resulted in the total depopulation of at least one of New Zealand's island sanctuaries.

Probably the single most revealing document on the necessarily secret business of trafficking in live animals is the autobiography of a French animal trader, Jean-Yves Domalain. In his book, *The Animal Connection: the Confessions of an Ex-Wild Animal Trafficker*, first published in 1975, Domalain gives a detailed account of the mostly illegal and extremely sordid business dealings of a typical animal trader. He confesses that, until 1970, he ran a widespread animal trafficking business from South-east Asia. Amongst other exotic animals he smuggled gibbons, clouded leopards, black leopards, douc monkeys, and many species of birds. Although his activities were certainly illegal, he traded almost exclusively with 'legitimate' zoos and circuses. On average, Domalain estimated, 80% of the animals handled either died in transit or within the first few months of captivity.

One of the most remarkable revelations in this book is that, even among the most reputable dealers, 'legal' animals are rare exceptions. Domalain believes that more than 90% of the trade in live wild animals handled by dealers anywhere in the world is illegal. First-hand experience led him to conclude that it is virtually impossible for anybody connected with the wild animal slave trade to have clean hands.

To begin with, the hunters of wild species are supposed to be on the commercial register. They are required to have licences or permits, must observe restricted hunting seasons and must not hunt protected species. In Domalain's experience, none of these requirements were ever complied with. With the exception of the highly organized commercial companies in East Africa which practise relatively humane methods of capture with large animals such as elephants, zebras and giraffes, animals are captured by peasant trappers without any legal observation. And the volume trapped this way is enormous. In 1988, the British-based Environmental Investigation Agency (EIA) estimated that ten million wild birds were exported from Senegal alone each year – and that at least that number die *before* export.

Although often excellent hunters, these peasant trappers are very poor at live capture. Their methods are generally extremely wasteful and often cruel. For example, most apes and monkeys are captured by hunters who shoot the females who are carrying young. Usually about one in four infant monkeys survives when its mother falls from the trees. These are taken captive. The majority of animals delivered to dealers are impossibly maimed – often crippled by snare wires, often with broken or missing limbs, starving, or near to death through disease or gangrene. These animals are summarily rejected and slaughtered for their skin, fur or feathers.

The trappers sell their animals to middlemen who are supposed to obtain appropriate authorization for export. Just such an operation was uncovered by EIA investigators in 1987 while looking into the smuggling of protected baboons from Gambia. This illegal smuggling has been organized for years by Leon Masfrand, a French Consul in Senegal. Masfrand acquires up to 1,000 primates a year and sells them to research clinics: the French Atomic Energy Commission and the French Natural Centre for Scientific Research are major clients. EIA found that baboons too large for research labs were simply strangled with a wire noose and discarded. Despite film and documented evidence, Masfrand still retains his post and his business. He is just one of thousands profiting by this shadowy trade. Middlemen, like the French Consul, are also supposed to observe international regulations concerning premises, shipping conditions, installations, quarantine quarters, isolation cages for carnivores and so on. In fact, animals are seldom vaccinated or checked by vets for epizootic diseases or rabies, etc. What regularly happens is that payments are made, and blank certificates handed over to the exporter who himself falsely certifies that these regulations have been complied with. Exporters' guarantees and the certification of animals in most Third World countries are generally regarded as a black joke. Consequently millions of animals contaminated with diseases – like salmonella and psittacosis (parrot-fever) – regularly result in a considerable number of human deaths each year.

If problems arise over the transportation of animals, outright bribery of airport and customs officials is the order of the day. In Bangkok, the going rate is between $10 and $100 a crate, depending on the animal. In Bogota and Nairobi charges are similar.

On the import side, in places such as Europe and North America – where bribery of officials is less common – fraudulent or stolen documents with official stamps and appropriate signatures are often used. Customs inspection, even at large airports in wealthy countries, is ludicrously inadequate. Frontier and airport officials can hardly ever tell the difference between one species of cat or monkey or bird and another. Furthermore the volume of trade is so high that it is impossible to check more than a fraction of the animals. Even in more obvious cases, cages with dual compartments are used, with legal species prominently displayed and illegal species hidden. (Not uncommonly, this has resulted in the accidental smothering of the rarer, hidden group.) Then too, some animals are dangerous. How many custom officials are going to check a sack of pythons or a tray of tarantulas in case a prohibited species is included?

Domalain found that impressive letterheads from research institutions and zoos carried a lot of weight in the eyes of officials. Every trafficker has his own set of forged documents, rubber stamps and office equipment: official letterheads for zoos, rubber stamps for Customs & Excise, Department of Forestry, Ministry of Agriculture, veterinary surgeons and scientific research centres.

Domalain himself went to a stationer and overnight became head of the 'Laos Biological and Experimental Research Centre' as well as the – until then unknown – 'Vientiane Municipal Zoo'. He also awarded himself a doctorate in Natural Science and Veterinary Surgery. Many 'zoos' throughout the world, Domalain points out, are simply holding centres for animal smuggling operations.

Because of his impressive credentials, Domalain was given licence to export protected species at much higher prices than many of his colleagues. As he admitted in his book, notepaper with 'impressive letterhead (fraudulent), serial number of licence (spurious), permit (spurious), and trade register (spurious) and, most important, the number of an account with the Bank of Indochina (genuine), lent an air of authority to the whole thing'.

Fraud was also indulged in to improve the worth of many animals. Otters' fur was bleached with peroxide and labelled albino, thus increasing its value from $30 to $1,000. Common leopards became black leopards with the aid of a can of spray paint. (In some cases toxic paint was used, which proved fatal to the cats.) Many birds are dyed to improve the brightness of their feathers, or even to change them into other related species.

In the case of large mammals, and even some smaller ones, which are expensive to purchase but breed well in captivity, the animals' genitals are often damaged by exporters to make them sterile. As a result, importers will buy still more animals for breeding. The method of sterilization used on dromedaries involves crushing the testicles with clapboards, a mutilation almost impossible to detect. Other species require more complex methods, such as large dosages of ultraviolet rays.

What conclusions did the ex-trafficker Domalain reach? 'Do not buy any more animals, not one single animal more; don't go to the zoo, don't go to the circus. Otherwise, it will be goodbye to animals.'

In short, the trade in wild animals is a dirty business, and should be eliminated entirely.

4

The Trojan Toucan

Cocaine-stuffed birds from South America •
Marriage of drug and animal trade • Heroin and
Asian wild cats • Hashish and Afghan wolfhounds •
Drugs and trade in crocodiles, and snakes •
TRAFFIC: a conservationist's FBI?

Trafficking in illegal and endangered species is not only highly profitable but also loosely policed. If and when the traffickers are caught, there is seldom a prison sentence attached to the offence, and for the most part the fine is minimal and easily offset against profits.

Under such conditions, the trade in endangered species thrives. However, those involved in this ruthless trade – having already instigated systems by which their animal cargoes can get past customs officials – have seen opportunities to add a new, and even more profitable, element to their business.

Wildlife traffickers in South America and Asia have linked their businesses with traffickers in cocaine and heroin. The profits from this lethal combination are enormous. The operations manifest themselves in many and bizarre variations, but one common scenario is that of the 'Trojan toucan'. From Columbia, for instance, a shipment of several cages of live parrots and toucans is sent to the United States. Before shipment, two or three toucans are killed in each crate and stuffed with cocaine. The dead birds are then packed in with a dozen or more live birds and shipped. If inspected, customs officers see nothing unusual in this since it is fairly common for as many as half a consignment of birds to die in transit.

In the case of shipments from places like South-East Asia, the cages and boxes for some of the fiercer jungle cats were often made of laminated wood and heroin. This has proved to be a reliable ruse, as not many customs inspectors are keen to check the floorboards of a tiger's shipping crate.

In other parts of the world, other methods are used. Afghanistan, before the civil war, enjoyed a popular trade in suitably unfriendly Afghan wolfhounds in crates made of laminated hashish.

Probably the best all-round drug smugglers are the snakes. Indigestible packets of drugs are often force-fed to large pythons, then shipped. It takes considerable dedication in a customs officer to thoroughly examine a box of slithering, live snakes and check for evidence of drugs.

The International Union for Conservation of Nature, (IUCN), was established as a state-of-the-environment monitoring organization under the aegis of UNESCO in 1948. More recently it has found it necessary to set up a sub-organization, a kind of conservationist's FBI which monitors the wildlife trade. According to TRAFFIC, Bolivia and Paraguay are two of the worst offenders when it comes to animal trafficking going hand in hand with drug-smuggling. Manfred Niekisch of TRAFFIC's West Germany office claimed recently: 'We have proof that drug-smuggling and the illegal wildlife trade is controlled by the same Latin-American Mafia'.

Obdulio Menghi, an expert on South America's wildlife trade, found that in Bolivia the animal-drug combination was an open secret, and the many variations of the 'Trojan toucan' reveal a wealth of creative flair. 'It's cocaine in the wet season, and crocodiles in the dry,' joked one of the traffickers.

Some of these variations seem hardly veiled at all, as in the case of crocodile-skins rather than live crocodiles. The skins of crocodiles – actually they are spectacled caimen, much valued by French and Italian leather industries – are normally dusted with white preservative. One operation used cocaine instead of white preservative in alternate boxes, and when delivery was made at the European end the cocaine was simply vacuumed up. The skins went to one dealer; the cocaine to another.

The new element of drugs in the wildlife trade greatly increases the dangers involved in investigating and exposing the trafficking in endangered species today. The stakes are raised many times over, and the level of violence has rapidly escalated. In recent years there have been many individuals in the animal trade who have disappeared, or met a violent end as a result of tampering with the wrong toucan.

5

The Head-Hunters

Illegal trophy-hunting in US national parks • Pink elephants and Onassis' bar stools • Eskimo head hunters • Horn hunters of Africa and Asia: aphrodisiacs and status symbols • The rhinoceros and the Yemen dagger cult.

There is a peculiar tribe of head-hunters who for the past two centuries and more have wandered the world wilfully slaughtering and decapitating their victims. These are the trophy-hunters, the big-game hunters who call themselves 'sportsmen' and kill animals simply in order to have the heads removed and mounted on living-room walls.

Although not as popular an obsession as it once was, this head-hunting mania does continue, and is now spreading at an alarming rate to the newly wealthy in Third World Countries. There is a worldwide illegal industry catering to those who wish to acquire the heads of rare, and consequently endangered, species for their collections. Such hunters have already been responsible for the extinction of the Barbary golden lion, the Cape black-maned lion, the Bali tiger, the Arizona jaguar, the Atlas bear, the Mexican silver grizzly, the South African blue buck, the bubal hartebeest, the badlands bighorn sheep, and the Pyranean and Portuguese ibex.

Caring nothing for the lessons of history, and usually protected by their wealth, these hunters continue to kill animals for 'sport' even in the most policed and protected of reserves. In October 1973, for instance, US Fish and Wildlife Service undercover agents exposed an illegal ring of trophy hunters that charged its clients $7,500 – $20,000 for guaranteed kills of protected Rocky Mountain bighorn and other wild mountain sheep. The ring was run by a Denver police detective and a Montana minister, who organized hunts for endangered species out of season and within protected national wildlife preserves.

This ring was typical of the kind of illegal trophy-hunting operation that in America alone is worth tens of millions of dollars annually. The absurdities of some trophy-hunters are legendary. Wildlife illustrator David Shepherd relates the true story of the American trophy-hunter who had a whole elephant stuffed, painted pink and transported to his New York apartment where it stands next to his bar. In even worse taste, Aristotle Onassis had *his* bar aboard his yacht, the *Christina*, adorned with bar stools made from the penises of sperm whales. A less typical, but equally profitable group of hunters during the last fifteen years have been operating in a specialized market in the Arctic.

Until the early 1970s, the Eskimo people of Alaska were hunting about fifteen hundred walruses a year for their meat, skin and ivory, as they had done for centuries. Suddenly the slaughter escalated and became so severe that the Soviet Union made an official protest at the number of headless walrus corpses being washed up on its shores. The reason for the increase was the demand for trophies in the form of walrus tusks. The result was the killing of ten thousand or more walruses each year simply for the tusks. The wastefulness is considerable: a thirty-year-old walrus, weighing over a ton, is killed for the sake of its tusks which weigh only about sixteen pounds.

The scale and seriousness of the issue can be demonstrated by the evidence of a couple of exposed operations. In February 1981, the US Fish and Wildlife Service seized more than five tons of fresh walrus ivory valued at over half a million dollars. The same year other investigations caught a group of traffickers who had updated the tradition of addicting the natives to whisky and then trading it at inflated prices for their goods. This group were doing a direct swap with the Eskimos of tusks for cocaine.

There are, of course, a wide variety of trophy-hunters, some of whom seem stranger than others to Western nations. However, the truth is that almost any animal with tusks or horns has considerable difficulty in keeping its head these days.

In the Orient wild deer with horns of any size will be hunted to provide powders reputed to have many magical and medical properties, most of which are related to male sexual potency. The beautiful Schomburgk's deer of South-East Asia was literally hunted to extinction because of the demand for the imagined powers of its large multi-tined antlers. An absurd reason, some would say, for extinguishing a species, but no more so than that which caused the American eastern elk to become extinct: the demand for its upper canine teeth as a watchchain insignia for members of the Fraternal Order of Elks resulted in the extinction of the elk.

The most valuable single trophy-hunting market in recent years has been that of the rhinoceros. However, it is no longer big-game trophy hunters who pose the real threat to the existence of this second largest of all land animals. Its fate is now in the hands of a specialized market, virtually unknown to the world at large until the late 1970s, that is pushing all five rhinoceros species quickly towards extinction.

This market has grown out of a unique tradition among the Arabs of South Yemen, where the young sons of the wealthiest aristocrats are presented with ceremonial daggers made with rhinoceros-horn handles. Once such daggers, called *jambias*, were only available to a handful of aristocrats, but with the influx of great wealth from the Saudi oilfields, every Yemeni youth whose family can afford it receives a rhino-horn dagger as a status symbol. Consequently rhinoceros horn has soared astronomically – a single horn is now reputed to be selling at more than $24,000.

Yemeni daggers were not the only market for rhinoceros horn. The Chinese pseudo-pharmaceutical market in aphrodisiacs willingly pays over $600 a kilo for powder ground from rhino-horn shavings, while powder from the almost extinct single-horned Asian rhino is valued at $6,500 a kilo. On this market, whole Asian rhino horns have sold for as much as $40,000. Today, Taiwan has become the centre of the rhino horn market. Its close links with South Africa have allowed a smugglers' pipeline to be established.

Such markets cannot fail to attract hunters and cause an already seriously endangered species to be slaughtered by the thousands. All five existing species are on the edge of extinction. Both the black and the white rhinos of Africa have fallen well below twenty thousand, while the Great Indian rhino numbers less than one thousand. The forest-dwelling Sumatran rhino's population is less than two hundred, while there are only about fifty of the small, hairy, Javan rhinos left in the world.

Rugged to the point of appearing almost invincible, the rhinoceros in its armour-like skin has survived seventy million years of evolution, and a million or more years of human predation. However, it remains to be seen if it can survive a couple of decades of free enterprise trade in what amounts to a high-priced fad-and-fashion market.

6

The Ivory War

Ivory and slave traders of 19th century • Mass poisoning in 20th-century Zaïre • Armies of poachers with helicopters and automatic weapons • Elephant population declines from ten to one million • 'Kalashnikov Revolution': guns for ivory • Ivory 'laundry' operations • 80% of all ivory illegal • Call for international ban.

'Ivory has been steeped and dyed in blood. Every pound weight has cost the life of a man, woman or child; for every five pounds, a hut has been burned, for every two tusks a whole village has been destroyed, every twenty tusks have been obtained at the price of a district with all its people and plantations.'

This statement was made in 1890 by the famous explorer, H. M. Stanley, in his book *In Darkest Africa*. It was no exaggeration. The searchers for ivory in Africa in the eighteenth and nineteenth centuries caused misery and death on a massive scale. Turning European guns against African tribes in order to expand the range and profits of ivory-hunting, the traffickers soon combined the gathering of 'white gold' with that of 'black gold' – the slave trade.

Sir Henry Stanley and most other early African explorers reported numerous encounters with the ivory and slave traders. In one such encounter in central Africa, Stanley reported with disgust a train of some 2,300 women and children chained together. All who were large enough to serve as porters were carrying elephant tusks. He also learned that all the menfolk had been put to death and that five earlier expeditions had yielded similar results that year.

The human toll in the ivory trade, although still considerable, has been much reduced since that occasion witnessed by Stanley.

However, the toll so far as elephants are concerned is much worse, and that legendary ruthlessness of the ivory traders towards both man and beast resurfaces whenever the possibility of a profit is sighted.

In 1978, on the Zaïre river near Kisangani, once named Stanleyville after the great explorer, a gang of elephant poachers organized a massacre of particular ingenuity. The aim of the poachers was to kill as many as possible of the elephants in the region at a single stroke, and harvest their tusks. However, rather than shooting or trapping the animals, a cheaper and simpler means was planned. In collusion with corrupt government officials, the elephant poachers dumped twenty-two tons of lethal pesticide into animal waterholes along the riverbank.

There is no argument as to the plan's success. The poison wiped out thousands of elephants: the entire population, right down to nursing calves. The poachers made their fortune cutting the tusks out of the dead animals. However, as they must have foreseen, the pesticide also killed tens of thousands of other animals who used the watering holes. Furthermore, it killed, or poisoned to a crippling degree, many of the animals and indeed many humans who attempted to feed on the meat of the dead or partially poisoned animals.

Hundreds of wardens and others have died during the last forty years of the Ivory War in Africa, and as the price of ivory rises, so does the murder rate and the level of corruption in the ranks of the government and the military. Even when traffickers are caught, there is no guarantee that justice will be carried out. In 1978, a notorious gang of thieves was caught red-handed with over $100,000 worth of raw ivory in Luangwa National Park. In the previous seven years, ten game wardens had been murdered by poachers in this park, yet the Zambian court – which could have ordered fines of up to $40,000 each and five-year prison sentences – decided to impose derisory penalties. Each poacher was fined only $1,000, and the judge ordered that the poachers' guns, which had been confiscated by the park wardens, were to be returned to them!

In 1870, the estimated elephant population of Africa was ten million. A century of ruthless ivory trading brought that number down to five million by 1970. Between 1970 and 1980, however, the rate of killing escalated at an unprecedented rate. By 1980, the elephant population stood at just 1,300,000.

During this decade of indiscriminate slaughter, as we have seen, Idi Amin in Uganda, and the 'Emperor' Bokassa in the Central African

Republic organized the virtual extermination of elephants in their countries. Similar scenarios were seen throughout Africa, as corrupt politicians became rich and businessmen, unable to export currency or trade in gold under national monetary restrictions, resorted to trafficking in ivory as a hedge against inflation, and a means of stockpiling a source of wealth outside countries with unstable governments and currencies.

During the 1970s Zaïre became the biggest single source of ivory, even though its export was officially banned in 1976 by President Mobutu. Investigations revealed that the ban had little or no effect on poaching and smuggling operations. Four members of the ruling political bureau and one member of the president's own family were involved in the nation's ivory trafficking business. It was an open secret that C-130 cargo planes were being flown with massive monthly ivory shipments to South Africa.

It may come as a surprise to the average purchaser of ivory sculpture or jewellery to hear that the chances of purchasing legal ivory in even the most reputable of shops are slim. At least 80% of all marketed ivory is illegal. Even the most cursory examination of documents pertaining to the ivory trade reveals that 60% – 90% of all ivory is exported with fraudulent or illegally obtained papers. Furthermore, not one African nation exports ivory in conformity with the international treaty documentation set up for its trade, although all were signatories of this agreement.

Since 1980, a big push has been launched in the ivory war against the last wild elephants in Africa. In what elephant expert Iain Douglas-Hamilton has called the 'Kalashnikov Revolution', ivory has become a currency to finance left-wing revolutionary and right-wing military causes throughout Africa. In some cases, ivory is directly exchanged for guns and ammunition.

By 1980, the Sudan had become the major clearing house for illegal ivory for the African continent. Thousands of tons of it have been shipped with 'legal' Sudanese government permits obtained by outright bribery. Sudanese ivory was technically legal, because according to its documentation it all came from legally killed Sudanese elephants. This was clearly ludicrous, for Sudan was an enormous 'laundry operation' for illegal ivory. Annual export figures indicate animal kills many times greater than the entire elephant population of the Sudan. In fact, less than 10% of the exported ivory originated in the Sudan. The rest was smuggled in, often over thousands of miles, by traffickers who raided protected national parks throughout the continent.

By 1988, the London-based Environmental Investigation Agency had called for an international ban in the ivory trade. The EIA had found that the elephant population had crashed once again from 1,300,000 in late 1980 to 750,000 in early 1987. By 1989 their estimate of the elephant population was less than 500,000.

EIA believes one major contribution to the rapid decline of the elephant is the CITES ivory quota system set up to 'control' the trade. This system – which is not based upon scientific management – gives a false legitimacy to the trade and enables the trading of poached ivory, making up 80% of the total, by various laundering techniques.

The EIA's undercover agents followed poached ivory from Burundi, Kenya, Sudan, Tanzania and other African countries through to Dubai and Singapore and on to the Far East. The ivory trail inevitably led to Hong Kong, the world's main market place where, usually in carved form, it is sold to Japan, Europe and America. Once poached ivory has entered Hong Kong it has reached base, and can continue on its way as 'legitimate' ivory.

In 1989, Tanzania became the first African country to call for an international ban on the ivory trade. This followed a smuggling investigation in which over 200 tusks were found in a container owned by the departing Indonesian ambassador. Similar seizures involved Iranian, Pakistani and Chinese diplomats, as well as a number of Tanzanian officials. This, the Tanzanian Wildlife Conservation society claimed, was simply 'the tip of the iceberg'.

The 'legal' system is so full of loopholes that powerful and wealthy dealers circumvent CITES restrictions with contemptuous ease. EIA's detailed work has shown that this quota system is unworkable and should be scrapped. Any delay in enforcing a world ban will prove fatal to the last of Africa's wild elephant herds.

As Stanley observed a century ago, ivory is the cursed treasure of Africa. The hand that seizes it becomes stained with the blood of both man and beast. Ivory is now a currency in itself. Throughout Africa it has become the medium of personal advancement and wealth for officials at every level in the government: military, judiciary and police. These corrupt men and women, for no reason but their own personal gain, are looting the heritage of all the African people.

7

The Whale War

Biggest of the big game animals: the whales •
Russian front: Greenpeace *vs* whalers • Harpoon
fired over protesters' heads • The pirate whaler
Sierra rammed then bombed • Spanish pirate
ships *Ibsa I* and *Ibsa II* bombed • Eco-espionage:
pirate ships seized in South Africa and Taiwan •
Whale as emblematic animal • 1972 UN moritorium
• 1985 IWC ban on whaling • Reason for failure:
corrupt Japanese politics • A choice.

The war at sea began in 1975, some two hundred miles off the coast of
California. There, the massive ten-storey-high, 750-foot Soviet factory
ship, *Dalniy Vostok*, and her fleet of six harpoon boats were gathered
for the hunt. For as the outlaw armies of traffickers in the wilderness
lands of Asia, Africa and the Americas closed in on the last of the apes,
the big cats, the rhinos, the elephants, and scores of other species,
these gunboat navies were at work on the high seas slaughtering the
last of the real 'big game' of the planet: the great whales.

Each Russian harpoon ship was 150 feet in length and armed with a
90-millimetre cannon loaded with a 160-pound exploding grenade
harpoon with foot-long barbed hooks. On 27 June, the harpoon ship
Vlastny picked up a herd of twelve sperm whales on her radar and
began pursuit. But just as the gunner on the high steel prow of the
Vlastny had the sleek backs of the whales within sight, a totally
unexpected factor was introduced to the hunt. Between the harpooner
and the whales, there suddenly appeared three inflatable zodiac
speedboats. Each boat held two environmental activists who, in order
to save the whales, were placing a human barrier between the gun
and the whale. They were gambling with their lives that a concern
for human well-being might save a few whales where concern for the
survival of the species itself had failed.

They lost the bet and the two activists nearly lost their lives as well.
The Russian gunner fired the cannon and the harpoon hurtled directly
over the middle zodiac as it dipped in the trough of an ocean swell.

With deadly accuracy, it struck the whale. There was an explosion of spray and foam and flesh; then a whirlpool of boiling blood. The steel harpoon cable came slicing across the water like a guillotine, barely an arm's length from the wildly veering zodiac.

This was the first direct-action gladiatorial on the high seas between whalers and conservationists. The activists were from a then-obscure Canadian ecology group called Greenpeace. And although their action did not save that particular whale, it had a great deal to do with saving many others, for film footage of the event widely publicized the conflict.

It was the first time the save-the-whale movement became front-page news, the first time the vague rumours of this ecology war at sea were crystallized in vivid images. But there was no doubt about the dangers involved. As a Greenpeace spokesman said: 'We were faced with the fact that next time we set out to block a harpoon shot, there would be no kidding ourselves that the Russians were unwilling to take a chance on killing us.'

They persisted, of course, and the zodiacs became the standard ecological floating pickets used to block whalers' harpoons. Sometimes there were variations: blockades, ship-boardings, and even activists chaining themselves to the harpoon guns. Over the next few years, Greenpeace and other action groups staged dozens of direct protests against the Russians, the Japanese, the Australians, the Icelanders and the Spanish. More harpoons were fired, protesters arrested, zodiacs and ships seized and held, but still the conservationists fought on.

These flamboyant actions, combined with more traditional methods, such as public rallies, petitions, publicity campaigns and political lobbying, were resulting in considerable success in the war to save the whale. However, in 1979, some environmentalists encountered a new and dangerous element in the industry that operated outside the existing controls on those who killed the whales.

This was the pirate whalers: gangs of modern freebooters armed with harpoon cannons and the whale-butchers' long flensing blades. Usually illegally set up and financed by the Japanese through dummy corporations, they acted very much in the swashbuckling tradition of the rough old sea pirates, violating all controls and regulations on whaling.

The most notorious of all the pirate whalers was the *Sierra*. It was a high-speed diesel harpoon boat that in 1968 had been converted into a combination harpoon factory ship with a rear slipway for slaughtering and storage freezers below decks. Ruled an illegal ship by the International Whaling Commission, this 'whaling-fleet-in-one' vessel zig-zagged across the Atlantic hunting endangered and protected humpback, blue and right whales. It also took undersized whales, nursing mothers and calves. It observed no closed season or protected areas, and found it could maximize its profits by taking only prime quality tail meat and dumping the remaining 80% of the whale carcass into the sea.

The ship avoided prosecutions by changing its name three times and shifting its nominal ownership from the Netherlands to Norway, Liechtenstein, Bahama, South Africa and Panama. To confuse matters further, it continually shifted its base of operations and went through a stunning range of flags-of-convenience: Holland, Bahama, Sierra Leone, Somalia, Cyprus.

In its various adventures, the *Sierra* had twice been run aground, experienced one mutiny, and been involved in a number of gunboat chases. It lost several court actions and had twice suffered bankruptcy, yet, phoenix-like, it always managed to rise again. The key to the *Sierra's* resilience was the extreme profitability of its unrestricted operation, and the willingness of the Japanese to back it and pay well for black-market whale meat.

By 1979, ecological investigators discovered an almost unbelievable international conspiracy financed by the Japanese to circumvent all whaling commission controls and rulings. In South Africa, Spain, Portugal, Taiwan, Korea, Peru and Chile, pirate whalers, modelled on the *Sierra* and owned or contracted by the Japanese, were hunting the seas without restriction.

Pirate whaling had reached epidemic levels. Even the supposedly 'legal' whaling operations of Japan, Russia, Norway and Iceland were cheating on quota agreements and violating rulings from the International Whaling Commission, the whalers' own organization. Many activists felt that it was time to fight fire with fire, and this led to some of the most extreme actions ever committed by ecologists.

It was mid-July 1979 when the ship *Sea Shepherd*, commanded by ecological activist Paul Watson, crossed the Atlantic to Europe. This was not a goodwill tour; she was looking for trouble. A few hundred miles off the Iberian coast, she tracked down her foe: the notorious pirate whaler, the *Sierra*. For a day and a night, it was a game of cat and mouse. The *Sea Shepherd* pursued and harried the whaler. On 16

July, both ships stood outside the Portuguese port of Leixos, near Oporto.

There, with what to the whalers must have seemed the fanaticism of a kamikaze, Watson deliberately turned his ship on the *Sierra* and, in a full-throttle attack, rammed the whaler. The *Sea Shepherd's* concrete-reinforced bow hit with the force of a huge steel axe. She ripped the whaler open with a gash eight feet long and six feet wide, and buckled the whole side of the ship.

Still that was not the end of the *Sierra*. Listing badly and taking on water, she did not sink. Quickly she managed to rev her engines fast forward and, lurching away from the berserk rage of the *Sea Shepherd*, she fled to the safety of the port.

Damage to the *Sierra* proved to be severe. It took nearly eight months for the pirate ship to recover from the hammering the *Sea Shepherd* had given her. But by the beginning of February 1980, she could be found quietly sitting at anchor in Lisbon harbour. Blessed with more lives than a proverbial cat, the *Sierra* was once again seaworthy, and ready to hunt whales again.

But the *Sea Shepherd's* crew were not the only hard men among the ecologists that the pirates had to contend with. Other eyes were watching the repair work on the *Sierra* in Lisbon, and early in February a team of specialists was called in.

At exactly 6.17 a.m. on 6 February, there was a loud blast in Lisbon harbour. The *Sierra* lurched in her mooring as a magnetic limpet-mine, placed six feet below the waterline, tore into the whaler. Into the ragged, gaping hole in her hull rushed the sea. In about ten minutes, the ship sank to the bottom of the harbour.

Less than three months after the bombing of the *Sierra* another Iberian port attracted the interest of the same extremist anti-whaling forces because it sheltered the little Spanish armada belonging to the notorious pirate whaler Juan Masso. This was the quiet harbour of Marin near the Vigo whaling station in the Pontevedro estuary.

On Sunday, 27 April 1980, at 2.00 p.m., two explosions rocked the harbour. The two Spanish whaling ships, the *Ibsa I* and the *Ibsa II*, shuddered at the dockside with the impact. No one was aboard the ships, or nearby, so by the time the crew reached the dock a few minutes after the explosion, less than ten feet of the upper mast was all that could be seen of either ship above the water, to mark where they sat at the bottom of the river-bed.

Most ecological activists profoundly disagreed with these extreme tactics, but they were united in the desire to stop pirate operations. To this end, the exposure of secret operations by ecological detectives in some cases proved as fatal to the pirates as the bombs.

Certainly this was true in South Africa, where in the same month that the Spanish ships were bombed, ecological investigator Nick Carter discovered two old whale-catchers being converted into new *Sierra*-style killer factory-ships. Carter's investigations revealed that the interests behind the *Sierra* were also financing these new ships, the *Theresa* and *Susan*. Through public exposure and diplomatic and legal pressure, Carter and others at last managed to force the government of South Africa to act decisively against the pirates. The two ships were seized and held. Then, finally, after a prolonged struggle for their possession, both ships were towed out into the Indian Ocean and shot to pieces by the South African navy.

The exposure of pirate ships in South Africa was followed by similar revelations in Peru, Chile, Brazil and the Philippines. But the effectiveness of ecological detective work in the right situation was above all proved by the investigations of Campbell Plowden and four other Greenpeace activists when moving against pirates in the South China Sea. Plowden revealed, despite official Taiwanese denials, that a fleet of not less than four pirate factory-ships was operating out of the port of Kaohsiung in Taiwan. As usual, Japanese money through dummy companies had set up the operation, and whalemeat was being smuggled into Japan through a complicated laundering operation which went from Taiwan to Korea to Japan to avoid detection.

Plowden's careful documentation of the smuggling and illegal whaling operation, through tracing the pipeline from the South China Sea to Taiwan to Korea to Japanese market traders, created a scandal in Japan and America that resulted in rapid sanctions being threatened against Taiwan for harbouring the pirate whaling and smuggling operation. Taiwan performed an act of rapid surgery. All four pirate whaling ships were seized and their whaling equipment was impounded. By the end of 1980, there was no more whaling in Taiwan.

The whale has become the emblematic animal for the entire ecology movement, the half-ton heart of the blue whale, the heart of the movement. For if this astounding animal, the largest on the planet, cannot be saved from extinction, what chance have lesser species?

In the course of the whale war, there is evidence that a revolution has taken place. More than all the ecology activists and their supporters, it has been the whales themselves who have been responsible for this: by virtue of their impossible hugeness, their gentleness, their evident intelligence, and the vocalizations that we now call whale songs. Many people now perceive a mystical aspect to these animals that have survived twenty or thirty times longer on this planet than man.

Whales are the Grand Canyon and Mount Everest of the animal world. The blue whale weighs as much as two thousand men. The brain of the sperm whale, at twenty pounds, is six times the size of the human brain. The bowhead whale has a mouth as wide as a car ferry's ramp, which would allow two semi-trailer trucks to enter side by side. The humpback whale sings strange and eerie arias over the open sea. The grey whale has a 7,000-mile migration pattern – the longest of any mammal on earth – ranging from the Arctic waters of Alaska and Siberia to the Baja Gulf in Mexico. The killer whale propels itself beyond the explanation of science at speeds of forty miles per hour and is the fastest animal in the sea.

Through the recordings of 'whale songs' we can hear the music of other worlds within our own, and perhaps this will help to guide us towards a revolution in the human mind, one that will recognize that we are but one of many beings who are meant to live in harmony on this planet.

Since the whale war's blitzkrieg of 1979–80, there have been a number of other direct action offensives, such as the *Sea Shepherd* and Greenpeace invasions of the Siberian whaling grounds in 1981 and 1982, and the devastating 1987 attack by *Sea Shepherd* agents in Iceland, in which the eco-saboteurs, Rod Coronado and David Howard, successfully scuttled two Icelandic whaling ships and demolished the country's only whaling station.

However, despite the dramatic nature of the gladiatorials between ecologists and whalers on the high seas, many of the most vital battles in· the whale war have been, and are still being, waged around conference negotiating tables, in government legislatures, and in courtrooms. High points for the environmentalists were the unanimous United Nations call for a ban on commercial whaling that, in 1972, gave a platform for the whole movement; the 1979 US Senate Packwood-Magnusson Amendment, which gave some threat of sanctions against violators of whaling regulations; and, most astonishing of all, the winning in 1982 of a two-thirds majority vote in the International Whaling Commission (IWC) to declare a moratorium on commercial whaling.

The save-the-whale movement has achieved some astonishing successes, and has certainly saved several whale populations from extinction. The Indian Ocean has been declared a whale sanctuary. Since the beginning of the whale war whaling has ceased in Australia, South Africa, Taiwan, Korea, Spain, Portugal and Brazil, among other countries. Pirate operations in at least twelve more have been forced to close down. Perhaps most remarkably, one of the last two giant whaling nations, the Soviet Union, retired its factory fleet in 1987.

Yet, in 1989, Japan, Norway, Iceland, the Faroes Islands and an unknown number of hidden pirate operations still hunt the whale. Considering the breadth and depth of worldwide support for the movement, it is the tenacity of the whalers that is remarkable. Following the 1982 moratorium, all treaty and legal agreements relating to the whaling industry – from the UN to the IWC – ruled that commercial whaling operations should cease by 1985. Since then, with the exception of aboriginal hunting, all whalers have been considered pirate operations. Yet, by January 1989, Greenpeace activists were having to go back on to the high seas. This time they were literally pursuing the whalers to the ends of the earth: for ten days the Greenpeace ship *Gondwana* blocked the Japanese whalers in the Antarctic Ocean. But whaling still goes on. Why?

There is only one reason. A handful of men are making large profits in virtually the only market for the product remaining in the world: Japan. All commercial whaling (and certainly all pirate whaling) has continued for the past twenty years simply to supply the Japanese market with whalemeat. This market, which buys and sells legal or smuggled meat indiscriminately, only survives itself through government subsidization and artificial pricing. In the end, the whale war continues because of extremely corrupt internal Japanese politics, and the pressure that a few wealthy fishing industrialists can bring to bear on the Japanese government.

An end will come to the whale war one way or another. In its time, the whalers have made eight of the ten great whale species commercially extinct – that is, sufficiently rare to make them not worth hunting. Do we continue to let them ply their trade until commercial extinction leads to biological extinction? It must be a clear choice. Do we extinguish the industry, or do we allow the industry to extinguish the whale?

3

Welfarists and Humaniacs

Campaigns against cruelty to animals

'A man is ethical only when life is sacred to him, that of plants and animals as well as that of his fellow men, and when he devotes himself helpfully to all that is in need of help.'
Albert Schweitzer

1

Humaniacs Versus Hunters

A modest proposal: Cleveland Amory's 'Hunt the Hunters' club • Anti-hunt alliances • Shock tactics • Bloodsports: fox hunts, bull fights, cock fights, • Hunting casualties.

On American television's NBC *Today Show* in the early seventies, there appeared a large frizzy-haired man reading a proposal for the creation of a new conservationist club. He suggested that the country currently had a problem with its hunting fraternity. There were about fifteen million hunters in America, and many concerned citizens had complained that on holiday weekends they were cluttering the countryside and highways of the nation. Paraphrasing the hunters' own argument for the control of populations of wildlife species, the gentleman suggested that 'an intelligent long-term programme of conservation of hunters' was needed to manage the alarming numbers.

To this end, he was announcing the formation of the Hunt the Hunters Hunt Club which boasted the motto: 'If you can't play a sport, shoot one'. It was not, he quickly pointed out, an attempt to exterminate hunters, but simply a culling in order to 'trim the herd'. 'It proposes a carefully regulated open season on hunters where you and your clubmates, in a carefully regulated gentlemanly club atmosphere, can have a really first-rate weekend shoot.'

The club would demand strict enforcement of rules: bow hunters to be shot with bows and arrows, fox hunters to be ridden down on horseback with only purebred dogs, trappers to be trapped – humanely, of course. Naturally there would be sportsmanlike controls and regulations. 'Please do not, for example, simply go out and take pot shots at hunters – within city limits, say, or in parked cars, or in their dating season.'

The club also expected its members to conduct themselves with tact and sensitivity. After bagging a hunter, for example, the club advised members against draping the corpse over the automobile in an ostentatious fashion. Members were also advised that 'mounting heads is considered by the club to be in very bad taste'. It was also suggested that just having the cap or jacket prominently displayed should suffice. When in doubt, 'just use your judgment and the inherent good taste of all sportsmen'.

Finally, the speaker mentioned the concern expressed by some club members conducting such a hunt about the reported toughness of hunters. This, he reassured viewers, was patently untrue. Properly prepared and seasoned, 'they can be quite tasty'. The frizzy-haired zealot who made this modest proposal for the Hunt the Hunters Hunt Club was Cleveland Amory, one of the more flamboyant spokesmen for those who deal with the issue of animal cruelty. Amory is the head of the American 'Fund For Animals' which has recruited scores of celebrities and tens of thousands of citizens to the cause.

An outspoken advocate of 'the right to arm bears', Amory is a no-compromise enemy of hunters, trappers, vivisectors and just about anybody who harms or victimizes animals. He was once asked by a reporter for a word of compassion for two hunters who had been clinging desperately for two nights to a mountain-ledge awaiting rescue. Amory demurred: 'I'm rooting for the ledge.'

Amory is one of many who have taken a new, aggressive stance on issues of animal welfare. He belongs to a breed of animal welfare activists who, often only half-jokingly, refer to themselves as 'humaniacs'. Too often in the past, they believe animal welfare groups have been unfairly characterized as bleeding hearts, and 'little old ladies in tennis shoes'. Amory advocated 'putting cleats on the tennis shoes' and, further, has allied himself with some tough and intelligent activists, lobbyists and campaigners.

Not that the enemies of the humaniacs have been taking their radical actions lying down. Every move to introduce humane legislation in any area of animal welfare has been bitterly opposed. To the humaniac, making headway in areas of national 'blood sports' often appears to be impossible. Those who oppose, say, the cruelty of bullfighting or cockfighting in Spanish-speaking nations face a long struggle towards any sort of recognition. Even in Britain, where animal welfare concerns are relatively strong, institutionalized hunts are still considered honourable traditions that must be maintained at all costs.

Some fox-hunters, when confronted by those who oppose the hunt, have been notable for their extreme self-righteousness. In 1976, to give an example, a member of the Essex Union Hunt was indignant in his own defence when charged with savagely assaulting a hunt saboteur. Indeed, he seemed to live up to Oscar Wilde's definition of the foxhunt: 'the unspeakable in pursuit of the uneatable'. It was his contention that 'Horsewhipping a hunt saboteur is rather like wife-beating – they're both private matters'.

In America it is the hunters of deer, ducks and other 'game' animals who consider themselves inheritors of a national tradition. To this end there are numerous powerful lobbying organizations such as the National Rifle Association, which seems to have, as one of its objectives, the fighting of every piece of conservation or humane legislation that is proposed. Many of the humane lobbyists argue for the cessation of hunting on the considerable evidence of violence unnecessarily inflicted on some animals. However, there are others who oppose hunting for specifically human reasons.

The casualty figures for humans in hunting incidents are staggering. Every year in America alone, between five hundred and a thousand people are killed, and well over seven thousand are wounded in hunting accidents. There are also many highly questionable 'accidents'. In California in 1979, for instance, two hunters were successfully convicted of murder when it was learned that because they could not find any other game to hunt, they had intentionally stalked an unarmed hiker, and shot him dead.

The issue of animal cruelty covers such a broad spectrum that, inevitably, certain specific areas have been targeted and major campaigns launched in order to increase public awareness of it.

2

Ice Capades:
The Seal Hunt

Twenty-year war on ice • Cause célèbre of
humane groups since 1964 • Moral and economic
arguments • Uses of sealskins • Actions by IFAW,
FFA, Greenpeace • *Sea Shepherd*, Franz Weber,
Brigitte Bardot • Magdalen Island attack • EEC ban
and end of hunt • Loopholes.

From the mid-1960s to the early 1980s, the *cause célèbre* of the animal
welfare movement was the annual Canadian harp seal hunt. The
dramatic impact of documentary films and photographs of the seal
hunt, showing the bludgeoning of helpless four-to-twelve-day-old
whitecoat seal pups in hundreds of thousands, did much to mobilize
public opinion against brutality towards animals. In spite of
arguments about the economic needs of fishermen and claims of over-
population of seal species, the fate of the seal pups and the obvious
cruelty of the slaughter became a world-wide issue.

Since the first publicity provoked concern in 1964, the Canadian
government aligned itself with the sealing industry in censoring the
press and intimidating animal welfare agencies who wished to
monitor the hunt. However, the public outcry over photographic
evidence of baby seals being skinned alive in front of their mothers
forced the government, in 1966, to introduce the Seal Protection Act.
This legislation was reputedly drafted to protect seals in their habitat
and prevent inhumane slaughter.

From the time of the introduction of this law until 1977, when the
protest movement reached its peak, over two million whitecoat seal
pups were killed. Yet despite repeated documentation of violations of
the Act and the counter-argument of authorities that the seal hunt was
the most 'rigorously regulated' in the country, not one sealer was ever
prosecuted under it.

The fact was that the so-called Seal Protection Act was used solely as
a vehicle for arresting, controlling and intimidating those who really

wished to protect the seal. It was also used to intimidate and censor members of the press who wished to document the seal hunt. The absurd degree to which the government policed the hunt made it appear to be a major issue of national security. Indeed, most national military zones in Canada were far less protected.

Regulations were implemented so that photographing, or even witnessing, the hunt could lead to arrests and confiscations. Flying over or near a hunt area – even when the hunt was not in progress – could and did result in the seizure of aircraft and the arrest of all on board.

Patrick Moore, a Greenpeace director, was arrested under the articles of the Seal Protection Act because he covered a seal pup's body with his own in order to prevent it from being clubbed. Police dragged Moore away from the seal, and a sealer immediately rushed in and broke the pup's skull with his club. Moore was afterwards provoked to point out that under the so-called Seal 'Protection' Act, the only thing that wasn't illegal was hitting the animal on the head, and skinning it.

The tradition of the annual seal hunt in Newfoundland is an old one but, despite arguments to the contrary, it had long ceased to be a major source of income. Most of the profits in fact went to Norwegian seal-ship owners. Less than half a million dollars annually reached the 'landsmen', the rural Newfoundlanders who, the government argued, were the chief beneficiaries of the hunt. Of all landsmen licensed to kill seals in 1976, for instance, 60% earned less than $100 annually, and fewer than 8% earned more than $1,000.

The irony is that the Newfoundland landsmen have traditionally been nearly as badly exploited as the seals themselves by a handful of greedy ship-owners and merchants who controlled the sealing and fishing industries. For over two centuries they kept landsmen in a state of virtual serfdom by managing all aspects of their lives. They set their own prices for the fish and seal pelts they bought from the landsmen and sold them at great profit. And to maximize profits, they did not bother to maintain their sealing vessels properly: some owners actually removed safety devices such as lifeboats and radios in order to cut costs. Several major disasters were due entirely to criminal negligence on the part of ships' captains and owners, including the 1914 catastrophe when 255 sailors died on the ice.

By the 1950s there had been some improvements in safety standards, but the landsmen were still being misused by the ship-owners and merchants and were seeing very little of the profits from what was increasingly a marginal industry.

By the 1960s and 1970s, the argument that the seal hunt was economically justifiable had become ludicrous. One activist pointed out that the seal hunt's gross earnings in the Newfoundland economy were equal to the contribution made by two Macdonald hamburger chain restaurants. Furthermore, the federal and provincial governments were spending far more than this on public relations and policing the event. Added to this, the hunt was becoming a major embarrassment internationally, which was earning Canada a shameful reputation for barbarity.

By the mid-1970s it seemed as if every major welfare group was involved in direct action or observer roles out on the ice floes in order to keep the world informed of the scale of the slaughter. An early leader, and major champion of the cause, was Brian Davies. Davies was a New Brunswicker who formed the International Fund for Animal Welfare (IFAW). Through its publicity campaigns and provocative actions which led to arrests and imprisonment, media attention dramatically increased. Later both *Greenpeace* and *Sea Shepherd* activists were also in the front line, shielding the bodies of seal pups with their own, spraying harmless dye on the animals in order to make their pelts unmarketable, and generally interfering with the hunt. Many were arrested and roughly treated. Paul Watson, *Sea Shepherd's* commander, was almost drowned in icy water by irate sealers.

One of the more high-profile protesters was Brigitte Bardot, who added her name to the struggle and went out on the ice with the protesters. The sealers' response to the French actress proved far from gallant. In the midst of her press conference the bloody carcass of a newly skinned seal pup was hurled at her.

Franz Weber, the Swiss conservationist, offered to finance a million-dollar toy and synthetic fur factory in Newfoundland if, in exchange, the hunt was abandoned. Although this industry would have provided more jobs and been more profitable than the hunt, the offer was rejected out of hand.

The seal war escalated, with increasing numbers of arrests, confiscations and imprisonments each year. It reached a peak on the Magdalen Islands when an environmentalist's helicopter was deliberately destroyed. Police looked on passively while supporters of the seal hunt overturned the craft, then smashed it up. The mob then proceeded to beat and spray paint on the environmentalists and even some of the newsmen who accompanied them. Despite the fact that in this small island community, the police knew every citizen by sight, the only arrests made during the incident were of the environmentalists themselves. Veteran newsmen claimed that the scene was reminiscent of the civil rights marches in the Deep South during the early sixties.

Although national surveys showed that the majority of Canadians opposed the hunt, the structure of representative democracy in Canada is such that regional prejudices often outweigh a national consensus. This was certainly true of the seal hunt. The issue was politically isolated to those regions where it was carried on. Any federal or provincial politician in such areas who opposed the hunt was certain to be defeated at the polls and so felt compelled to support it, whereas elsewhere it was a non-issue that would not result in the loss or gain of a single seat.

It soon became apparent to environmental strategists that mass popular opinion was powerless to alter the status of the seal hunt in Canada. The only way to get results was by destroying the seal fur market itself. If there were no buyers for the pelts, they reasoned, there would soon be no reason to hunt.

Fortunately for the environmentalists, unlike the Japanese whalemeat industry, the Canadian sealskin market was not domestic. It was entirely dependent on foreign buyers and, consequently, it could be influenced by international public opinion.

The beginning of the end for the Canadian seal hunt came in 1972 when the US Marine Mammal Protection Act ended trade in whitecoat sealskins within that country. But the death knell was not sounded until 1982, when the British MEP, Stanley Johnson, presented a petition of three million signatures to the European parliament and shepherded the EEC ban on sealskins through to legislation. Following this, in 1983, Canada was faced with a threatened boycott of its fish over the seal hunt issue. The Fisheries Council of Canada, which had everything to lose, pleaded with the government not to risk a billion-dollar fish export market.

By 1984, the Canadian seal hunt was virtually finished and its markets destroyed. However, the government was not yet willing to admit defeat. It was not until 1988, long after the publicity had died down, and when ecologists' interests were directed elsewhere, that a Canadian government minister made a low-key announcement. The Canadian harp seal hunt, he declared reluctantly, was now officially ended and was unlikely ever to be revived.

It seemed as though the twenty-year Seal War was finally over, and the animal welfare activists had achieved their victory. Sadly, however, it now appears this victory had been a limited one. Using a strategy comparable to the Japanese in the whaling industry after the moritorium on commercial whaling, the Canadians have managed to find a way of hunting using a legal loophole. In 1988, Canadians actually killed 70,000 seals, but they avoided sanctions because they did not actually kill any 'white coats' (baby harp seals) or 'blue backs' (baby hooded seals). Although most of the slaughtered animals are legitimately adult seals, a considerable number are what are known as 'raggedy jackets' – white coats and blue coats which after fourteen to sixteen days shed their original coats. In other words, the sealers simply allowed the baby seals to survive for another week before slaughtering them.

Betting that the public is now sufficiently confused by the twenty-one year struggle, pro-hunt strategists believe there is a chance of re-building the industry. They are correct in that most people now believe the seal hunt is over. (Nor are they aware that besides the 70,000 killed in Canada, there were comparable numbers of seals killed in Greenland, Namibia, Norway and Russia which brought the worldwide seal kill in 1988 to over 300,000). However, the strategists are quite wrong if they believe the ecologists are going to quit now. In October 1988, the EEC anti-seal ban was again reinstated.

3

The Fur War

Annual kill of fur trade: 270 million domestic and ranched animals; 50 million wild animals • Target: leghold trap • Unintended victims; unacceptable cruelty • David Bailey's 'Dumb Animals' • Cleveland Amory's 'Real People Wear Fake Furs'.

The massive international press coverage given to the Canadian seal hunt in the late-1960s helped to fuel the controversy over the morality of the use of furs, which were still very much in fashion. By the mid-

1970s, not less than 320 million animals a year were being slaughtered to provide fur coats and other fur products.

Quite apart from its violation of national and international laws by marketing critically endangered species, environmentalists oppose the fashion fur industry because of the issue of animal cruelty.

The statistics given in Greta Nilsson's Animal Welfare Institute publication, *Facts About Furs* show that approximately 240 million domestic animals are killed annually for their pelts: rabbits, sheep, and Persian lambs (*karakul*). A further 30 million killed are wild 'ranched' animals, mostly mink and fox. The remaining 50 million animals slaughtered each year are entirely wild, fur-bearing animals hunted or trapped with varying degrees of cruelty in their own habitat.

Many people take issue with the entire fur trade, but the main thrust of the protest against the industry is directed at the 'wild' fur market. The major objection of animal welfare groups concerns the use of steel leghold traps. These traps, which secure but do not immediately kill the animal, usually result in a slow, painful death. The leghold trap in America alone kills approximately 10 million fur-bearing creatures annually. It also kills at least another 20 million unintended victims which accidentally step into its jaws: domestic dogs, cats, porcupines, squirrels, birds, etc.

The leghold trap has been banned in over fifty countries in Africa, Latin America and Europe. North America and Asia (with the exception of India) have held out. In the USA there is a powerful lobby of forty thousand professional trappers and about two million 'recreational' trappers that has, so far, blocked any move to ban these traps.

The activists' argument against the leghold trap is that it is both wasteful and cruel. Wasteful because it traps more than twice as many non-target animals as the intended prey. Cruel because it severely wounds and maims without immediately killing. Once trapped, and likely suffering a broken limb, the animal may remain that way for up to a week. If the animal does not succeed in releasing itself – as many do by desperately chewing off a trapped paw – it will die slowly from gangrene or hunger and thirst. If it survives the long wait, it can only hope for the mercy of a quick end when the trapper clubs or kicks it to death.

There are alternatives which are widely and successfully used, such as box traps and the swifter-killing Conibear trap. However, trapping lobbyists have steadfastly refused to change their methods.

There have been many protests at the fashion end of the fur trade in recent years. Some were individual actions such as the ritual burning of fur coats outside department stores which sell furs. Others organized consumer action boycotts against stores that carried wild furs, while still others picketed fur fashion shows.

Brigitte Bardot, who gave her support to the seal protest, and many other celebrities normally identified with high fashion circles, including Zsa Zsa Gabor and Doris Day, have stood up to be counted among those who oppose the wearing of furs, as has photographer David Bailey, who has filmed some of the most shocking of Britain's LYNX anti-fur campaign films. Usually dubbed the 'Dumb Animal' campaign, its effective slogan is: 'It took 40 dumb animals to make this coat, but only one to wear it'.

Years earlier Cleveland Amory's Fund For Animals had organized a vigorous advertising campaign to promote artificial furs in a hopeful bid to deter the fashion for real furs. The caption read: 'Real people wear fake furs'. The models in their elegant artificial fur coats included Mary Tyler Moore, Doris Day, Angie Dickenson, Jayne Meadows and Amanda Blake.

In 1988 anti-fur demonstrations were held in 70 American cities on 'Fur-Free Friday', the day after Thanksgiving. In New York, a march of several thousand anti-fur demonstrators was led down Fifth Avenue by the TV personality, Bob Barker. Barker had already made his position clear on this issue. Identified for many years as the traditional host of the Miss USA and Miss Universe pageants, Barker shocked organizers of the event by resigning in 1987 when officials insisted on presenting the winners with fur coats. Barker feels strongly that fur coats are inappropriate status symbols. 'It isn't necessary to torture and kill animals to show people how much money you have', he told reporters. 'You can buy a nice cloth coat and pin money to it'.

Barker is now among the scores of celebrities who endorse the Friends of the Earth call to ban all wild furs. As animal welfarists vividly point out: 'Any woman who wears a wild fur has on her back at least 150 hours of torture'.

4

Dolphin Raiders

US fishermen's dolphin kills • Japan's Iki Island
slaughter • Night raids by Dexter Cate and Patrick
Wall • Allan Thornton's undercover actions end the
Black Sea dolphin hunt.

America's mass killing of dolphins began around 1959. During the next
decade five million dolphins were slaughtered and America had
become the world's leading dolphin-killing nation, although it largely
remained a fisherman's secret.

This slaughter was brought about by an advance in the technology of
tuna nets and sonar systems which enabled the tracking of deepwater
yellow-fin tuna. No one quite understands the correlation between tuna
and dolphin, but it is known that schools of yellow-fin tuna often swim
deep beneath schools of dolphin. In spite of their proximity, these tuna
are virtually never found in the stomachs of dolphin, presumably because
the fish inhabit a level too deep for the dolphin to reach.

With the introduction of deep-sea nets, dolphins were trapped along
with the tuna and indiscriminately crushed and drowned. From 1959
onwards, this 'incidental kill' amounted to over half a million dolphins a
year.

By the late 1960s public awareness was growing of the dolphins'
needless slaughter. Groups involved in the anti-whaling movement
and the harp seal issue aligned themselves with dolphin protection
groups to fight for measures to be taken against fishermen who killed
dolphins. Despite the very powerful lobbying drive for exemption by
the tuna industry, the Marine Mammals Protection Act went into
effect in 1972.

The industry did not immediately succumb, but the pressure of
public opinion on government agencies resulted in the introduction of
new technology. A relatively simple device was found which allowed
dolphins to escape from fish-nets while the tuna remained inside. The
result was that by 1978 the half-million dolphin kills a year had been
reduced to less than twenty thousand.

Nor was this an end to the matter. American tuna purse seiners are now the best-observed vessels in the fishing industry, with inspectors frequently on board them. Furthermore, some fishing captains deserve considerable credit for their co-operation in reducing the incidental kill rate of dolphins dramatically during the last decade, even though a considerable number have attempted to avoid control by re-registering American ships in foreign parts. It is estimated that these vessels may still kill as many as 100,000 dolphins a year.

Even on American registered ships, there is not yet complete compliance with regulations governing dolphins. However, despite the restrictions of the campaign, and the stubborn resistance to change over the years by fishing interests, the dolphin campaign in America can be considered, at the very least, a limited victory.

Strong as the reaction was to incidental dolphin killings in America, it was mild compared to the furore caused by the deliberate slaughter of dolphins on Iki Island in Japan. Fishermen there, having obviously overfished their own waters, argued that the drop in the number of fish was due to the dolphins. Their solution was simple. They herded the dolphins that annually migrated past their island into a harbour and slaughtered them. Rather than discouraging this practice, the Japanese government turned it into a profitable pastime. Using the dolphins as a scapegoat for their poor management of fishing stocks, they ludicrously labelled them 'the gangsters of the sea' and fishermen were offered a bounty equivalent to $80 per dolphin killed.

The killing of dolphins on Iki Island was unique even in Japan, where there are large annual dolphin kills in other prefectures. For whereas in these other areas it could be argued that the animals were being used as food, on Iki Island, it was simply a programme of mass extermination.

As with the Canadian seal hunt, environmentalists presented the public with photographic evidence of blood-letting frenzy. Films showed the clumsy spectacle of men chopping and clubbing trapped dolphins while wading thigh-deep in blood-red water. There could be few who doubted that this was a sloppy and inhumane way to slaughter animals.

During 1979 and 1980 this annual extermination of dolphins on Iki Island attracted a good deal of international media coverage. However, little progress was made in restricting or controlling the kill until late one night in February 1981, in the midst of the annual slaughter, when a young American activist named Dexter Cate took a small inflatable kayak out to the island. There, in an enclosure in Katsumato harbour Cate found about three hundred dolphins thrashing around in the bloody water where hundreds of others had already been speared and clubbed to death during the day. Knowing that these would face the same fate at dawn, Cate went to the net floats and let down four nets, then gently herded the surviving dolphins out to the open sea.

Dexter Cate's action was the second sabotage of a Japanese dolphin hunt that year. A month earlier, Greenpeace activist Patrick Wall had released a hundred and fifty dolphins from the Futo harbour hunt in central Japan. Both men wished to make their acts of civil disobedience widely known in order to politicize the issue and so they allowed themselves to be taken into police custody. Cate's case in particular received wide public attention both in Japan and internationally. He spent three months in prison awaiting trial, then was given a six-month suspended sentence and deported.

Though Cate was ostensibly punished for his action, it would appear that the incident had an impact on the authorities. Not wanting further international censure over the issue, the Iki Island dolphin hunt was closed down the following year.

One of the most successful and efficiently engineered campaigns ever launched for protecting marine mammals was conducted in Turkey. Virtually unknown to the outside world until 1981, Turkey conducted a massive annual dolphin killing programme. Since the 1950s hundreds of thousands of Black Sea dolphins were slaughtered every year for conversion into chickenfeed and industrial lubricating oil. By the end of the sixties, the Black Sea dolphins had so declined in number that only Turkey continued to hunt them.

Despite the obvious dolphin population crash, by the end of the 1970s the Turks were continuing to kill 160,000 dolphins a year and, even after a further major population collapse in 1980, which resulted in only 54,000 dolphins being slaughtered, Turkey was still ranked as the number-one dolphin-killing nation.

In 1982 Canadian ecological agent Allan Thornton, under contract to the British-based People's Trust For Endangered Species, undertook an investigation of Turkish fishing ports on the Black Sea. At some considerable risk to himself Thornton conducted a campaign of undercover espionage in order to document and photograph the hunt.

Later that year, Thornton returned to Britain and published an exposé of the grisly slaughter, with colour photographs, in the *Sunday Express Magazine*. This, combined with representations to the Turkish ambassadors in London and Washington, achieved something of a small miracle.

By a stroke of good luck, the Turkish government was at that time negotiating a billion-dollar foreign-aid loan programme with the United States, and the Turkish authorities, not wishing the Black Sea dolphin hunt publicity to escalate to the level of the Canadian seal campaign, decided to end this nearly bankrupt minor industry rather than risk jeopardizing their hoped for loan.

Approximately one year from its inception, the Turkish dolphin campaign had been targeted, fought and won. By 1983, the Turkish hunt was over, and the Black Sea dolphins were a protected species.

5

Factory Farms

Cruelties of mass production • Issues and priorities • Target areas: battery chickens, intensive pig farming, milk-fed veal • Transport Legislation • Ritual killing for kosher and hallal meat.

Since the early 1950s, farming has undergone a revolution in methods. Nowhere is this more obvious than in the area of animal production which, in the last three decades, has seen innovations comparable to the rapid development of automobile assembly in the time of Henry Ford, resulting in the creation of the 'factory farm'.

The chief advantage of factory farming is high production rates utilizing minimal human labour and land resources. The disadvantages are the high costs of setting up and maintaining the factories, a high level of chemical pollution and the absence of concern for animal welfare. In an industry that in America slaughters nearly five billion animals annually, cruelty is not a cause for concern unless factory farmers become convinced that stress or illness is interfering with profitability.

Many animal welfare groups are pressing for reform in factory farm methods because they argue that these unnatural, crowded, intensive systems of animal production are inherently cruel. They also say that in most cases factory farming produces inferior products and these products are more susceptible to diseases on a mass scale (as Britain's recent massive salmonella-egg scandal certainly emphasized). One result of this is the increase in the consumer demand for such foods as 'free-range' chickens and eggs.

Some animal welfare groups maintain that vegetarianism is the only appropriate means of protesting against factory farming systems. Firstly, they argue that the slaughter of any animal is cruel and, secondly, that the production of meat is wasteful of grain which could be better used to feed millions of undernourished humans in the Third World.

Putting such issues as vegetarianism aside, however, there are three areas of factory farming that have drawn the most concentrated fire from animal welfare organizations: battery chicken operations, intensive pig farming and milk-fed veal production.

The mechanization of egg production in so-called battery chicken factory farms is now so great that some poultry farmers boast that only one worker is needed to service a plant of thirty to fifty thousand hens. One of the largest egg producers in America has a single factory farm with over two million hens divided into units of ninety thousand hens each.

Most visitors to battery chicken farms leave with the impression that they have just visited a kind of animal concentration camp. Battery hens are crowded into wire cages stacked ten deep, perhaps five to a cage measuring sixteen by eighteen inches or less. This gives each animal virtually no room for movement. They stand on wire floors where their eggs are laid. The eggs then roll on to conveyors beneath the cages, and are carried away.

Fed and watered by other conveyors, the hen lives out its one-to-two-year lifespan on an allotment of cage floor space equal to half a sheet of writing paper. In a building devoid of natural light, the only air it breathes is heavily polluted with ammonia fumes that come from tons of concentrated chicken dung. Not surprisingly, fighting often breaks out among the hens in such conditions. Rather than giving the animals more space, farmers attempt to cut down on the damage by 'de-beaking' the birds. Using a guillotine mechanism called a 'hot knife', they cut off the animal's beak. This painful operation is usually performed twice during a hen's lifetime.

Such living conditions are totally illegal under most animal protection laws for the keeping of birds. Virtually all countries with animal welfare laws specify that birds must be kept in cages of sufficient size to allow them to open their wings. However, so great is the power of the agriculture lobby that all such legislation contains an exemption for the poultry industry.

Many of the accusations of cruelty that are levelled against battery chicken farms have also been used against programmes of intensive pig and veal calf production.

Approximately half of America's pigs, for instance, are raised indoors on concrete floors in extremely crowded conditions, largely in the dark. The conditions of breeding sows are particularly unpleasant as they are kept perpetually pregnant or nursing in stanchions two feet wide and six feet long. The sows spend their entire lives in these steel and concrete stalls where they can sit or stand, but cannot walk or even turn round.

Milk-fed veal calves are a relatively recent innovation. Traditionally, veal calves were unwanted male animals produced in the dairy industry and usually slaughtered at a few weeks of age. However, the charge of cruelty was not levelled at calf production until the vogue for 'white veal', which originated in Holland, spread to America and the demand for veal increased. Now calves are allowed to grow to the age of thirteen to fifteen weeks, reaching over two hundred pounds, before being slaughtered. To produce a 'white' meat in larger calves, the beasts are tethered in stalls less than two feet wide and five feet long. The calves are fed a liquid diet of milk powder and nutrients to encourage rapid growth. They are also deprived of iron to make them anaemic and keep the flesh pale: a 'good' colour for veal. These animals are kept like chickens and pigs in conditions that allow them no movement at all, with poor ventilation and in almost total darkness until the day they are taken out into the light for the first time and killed.

Such is the way of life on the factory farm. The way of death is of equal concern to those in the animal welfare movement.

Long before the mass production methods of factory farming, animal welfare groups fought for the humane transport and slaughter of animals. In much of the Third World there is no legislation concerning the treatment of animals except that imposed by religious law. Even in

North America and Western Europe many questionable practices are ignored because of strong agricultural lobbies.

In America, for instance, the brutal excesses of animal shipping resulted in a ruling in 1906 that animals transported by rail had to be rested, fed and given water at least every thirty-six hours. There is no comparable legislation governing motor transport as this form of transport was not a factor in 1906; moreover, the farm lobby is so strong that every effort to close this loophole has been defeated. Consequently, it is not uncommon – and quite legal – for cattle to be shipped in trucks for thirty-six, forty-eight or even seventy-two hours without being unloaded. Because of this, an estimated 600,000 cattle die while being transported in America each year.

In North America and Europe, legislation governing humane slaughter stumbles badly when confronted by the dictates of certain religions. The major obstacles are the Jewish and Moslem methods of ritual killing which are observed in such a strict way as to preclude accepted ideas of humane slaughter. For example, the production of 'kosher' and 'hallal' meats requires the animal to be 'healthy and moving' before being killed by a single stroke of a sharp knife through the jugular vein and windpipe. This requirement has generally been interpreted in such a way that it prevents an animal from being knocked unconscious by either electric stunning or captive-bolt pistols, which are used by law in the majority of modern slaughterhouses.

The violence of ritual slaughter, as animal welfarists will point out, is not restricted to the slowness of death in bleeding from the throat. In most cases, cattle are shackled by one leg and lifted from the ground, upside-down, on to an overhead conveyor. When animals weighing over half a ton are jerked off their feet and hung by one leg, they inevitably experience a combination of muscle and ligament tearing, bone fracture, and joint dislocation. The animal, obviously in severe pain, hangs from the conveyor for two to five minutes before a clamp is inserted in to its nostrils to hold it still. Only then is the stroke of the knife delivered to its throat and its life allowed to ebb slowly away.

Interpretations of the requirements for ritual slaughter vary. In Sweden, for instance, kosher meat is not killed in this manner. Swedish Jews have accepted that ritual slaughter should not be exempt from the law and have consequently saved cattle in that nation considerable pain and suffering. In America and most of Europe, however, any effort to reform this procedure has been unfairly met with charges of racial and religious persecution.

6

Pet Traders and Petnappers

Domestic animals and the early welfarists • *Black Beauty* • 'Common' cruelties • Problem of exotic animals • Thirty-year tortoise war •Genetic cruelties • Pit dog fights • Petnappers: two million a year in US • Citizen and vigilante actions • Oregon Dognapping Trail.

When the first murmurings of animal welfarists were heard in the second half of the nineteenth century, their chief concerns were not unnaturally with the animals they knew best: domestic dogs, cats and horses. The publication of Anna Sewell's *Black Beauty* in 1877 was to the animal welfare movement what *Uncle Tom's Cabin* was to the anti-slavery movement. It proved a major catalyst in the passage of anti-cruelty laws governing horses, donkeys and mules.

As anyone concerned with animal welfare will readily tell you, in one horrifying story after another, cruelty to domestic animals – even to the extent of intentional torture – is still remarkably common. Furthermore, penalties for cruelty are seldom more than nominal fines, which, considering the extremity of suffering imposed on many animals, is hardly adequate. And although there are laws that can in some way censure intentional cruelty, unfortunately, a great deal of cruelty is also inflicted on animals through simple neglect or ignorance.

The pet trade in general is rife with merchants and consumers who are woefully ignorant about the correct feeding and environmental needs of animals. This is especially true of exotic birds, reptiles and mammals, less than 5% of which survive even a few years of captivity in the hands of generally inept traders and owners. Quite apart from the suffering inflicted on exotic animals taken from the wild, there is the added factor that a long-running fad for certain exotics has led to their extinction. This has already happened with many species of parrots, for example.

However, once a fashion starts, it is difficult to stop. Take the Mediterranean tortoise for example, approximately eight million of which were captured and shipped to Britain alone over a fifty-year period. Investigators found that of those tortoises that actually survived the journey, 80% sold through shops were dead within a year, and less than 1% of these long-lived animals survived five years of captivity. Not until 1954 did it become apparent that Mediterranean tortoises were rapidly diminishing in the wild, and it took the RSPCA and other organizations thirty years of continual campaigning before these tortoise imports were banned from the pet trade.

There are, of course, many in the pet trade who do know a great deal about animals, but some of these – such as animal breeders who practise what might be labelled 'genetic cruelty' – use their knowledge in cruel ways in order to make a profit. They breed all kinds of animals – everything from dogs to goldfish – with genetic faults that make them a novelty. Some are bizarre variations of common pets, such as so-called 'tumbler' pigeons and 'waltzing' mice. Lacking a sense of balance because of the inner ear faults for which they have been bred, most of these animals – with their almost spastic locomotion – eventually break limbs or severely damage themselves in other ways, and finally have to be destroyed.

Other cruelties inflicted through genetics are often less extreme, but more common. Surely there is questionable morality in creating artificial domestic breeds known to have weaknesses which cause suffering. The celebrated British bulldog is one of many examples. This animal has been bred to have a head so large that it cannot be born naturally, but must be delivered with the aid of a vet's knife. Although the bulldog is bred for its distinctively rumpled face, the folds of skin over its muzzle cause breathing difficulties which add to its usually chronic congenital respiratory problems.

There are also those in the pet trade who breed perfectly healthy animals for extremely cruel purposes. A current revival in illegal pit dog fights has resulted in the breeding and training of the tough and aggressive pit bulls – for no other purpose than to fight and kill other dogs. Pit dogfights are a gambler's game, and notable for their violence: the fight ends with the death of one of the dogs – and the winner is frequently so badly maimed that it too must be destroyed. Nor does the violence stop with the dogs themselves; numerous humans have been attacked and killed by these aggressively bred and trained animals.

If *Black Beauty* drew the public's attention to the cruelties exacted on horses, another book – Jack London's *Call of the Wild* – was equally revealing on the subject of cruelty to dogs. It also drew attention to the fact that even in the last century there was a black market in kidnapped family pets. As Jack London pointed out, when the Klondike gold rush opened up the market that suddenly required a huge supply of sled dogs, thousands of family pets disappeared overnight throughout America.

In 19th-century Britain, petnapping was a fairly common form of low-level extortion practised on wealthy families with pampered dogs. Petty criminals also often snatched the pets and held them until a ransom was paid – and in most cases the dog was returned.

Although hostage-style ransoming of pets seems to have fallen out of fashion, 'petnapping' has increased dramatically in recent years. In America alone, an estimated two million dogs and cats are stolen each year, and unfortunately their fate is often much worse than that of the hostages of the last century. Rather than being sold back to their owners, petnappers work with unscrupulous animal dealers. Cats and dogs are usually captured in one area, then driven in closed vans to dealers several hundred miles away. There they are held and eventually supplied to animal research establishments for experimentation. It can be a profitable business.

In response to this particular threat to domestic pets, which is often run in a highly organized manner, a number of angry citizens' groups have been established in America in recent years. One such group was formed in Virginia and called itself 'Action 81' because of the uncommonly high level of petnapping occurring along the Route 81 highway that passes through Virginia. Petnapping has become so common that Action 81 and many other organizations, like the Humane Society, which has branches in over forty states, now recommend that pet owners take the precautionary measure of having their pets tattooed with the owner's social security number. Such a marking would, at least, make it difficult for dealers to sell an animal to any reputable research institution.

Several animal vigilante groups have also been formed in Europe and North America in recent years. One such group is the Northern Animal Liberation League in Britain. This group of activists staged a raid on a Lincolnshire kennel that was known to be supplying dogs to laboratories for experimental purposes, and was suspected of involvement in petnapping. Proof of petnapping was confirmed when one of the dogs that had been taken from the kennels was positively identified and reclaimed by his original owner.

Despite claims to the contrary, it is not just small-scale animal dealers who are involved in the petnapping business. On 17 June 1987, Oregon's largest animal dealer, James Hickey, was found guilty of receiving stolen dogs. Hickey was selling nearly 500 dogs and cats a year to research laboratories. He was also found guilty of 27 violations of the humane provisions of the Animal Welfare Act and 30 record-keeping violations. Government inspectors found the animals in Hickey's care had been subjected to extreme cruelty and deprivation. Judge Victor Palmer claimed Hickey had violated the act 'deliberately, willfully and cruelly for personal gain and profit'.

In the severest penalty ever imposed for violations of the Animal Welfare Act, Hickey was fined $40,000 and ordered to suspend operations for 25 years.

7

'Humans Come First'

Anti-human charge • History from 1850: Animal Welfare Act first used to protect children • Animal welfarists active anti-slavers, anti-child labourists, and women's suffragists • Henry Salt's *Animal Rights* (1892); G. B. Shaw, G. K. Chesterton, William Morris, Gandhi • Albert Schweitzer's 'Reverence For Life'.

Inevitably, in any argument on the issues of animal welfare there is a danger that those concerned with animal cruelty will be accused of putting animals before people. Why this should be the case is something that often puzzles animal welfarists. In fact, anyone who considers the founders of the animal welfare movement will find that quite the opposite is true.

In 1870, a pioneer of the movement, Henry Bergh, managed to draft and push through the first piece of animal protection legislation in New

York to prevent the beating of small animals. However, it should be noted that Bergh's first use of that legislation was the prosecution of the custodian of a child for cruelty to a 'small animal'. Until Bergh's law was passed, cruelty to children was not an offence. This prosecution, and others which followed, led to the eventual formation of New York's SPCC – the Society for the Prevention of Cruelty to Children.

In Britain, founding members of the Royal Society for the Prevention of Cruelty to Animals (RSPCA) were William Wilberforce and Fowell Buxton, two of the greatest and most successful activists against human slavery in the British Empire. It was this same organization which, like its counterpart in America, established the British NSPCC – the National Society for the Prevention of Cruelty to Children. Moreover, it was Lord Shaftesbury, a leading advocate of animal welfare and an opponent of uncontrolled animal experimentation, who was a founding member of NSPCC and the initiator of the Factory Act which put an end to child labour and the fourteen-hour work day.

One of today's pre-eminent animal rights advocates, philosopher Peter Singer, credits his predecessor Henry Salt (author of the book *Animal Rights*, published in 1892) with anticipating most of his own concerns. Singer points out that Salt was a founder of the Humanitarian League, a group that worked for prison reform, against the flogging of children, and against vivisection and hunting. It encouraged pacifism and civil disobedience as a means to achieve social change. This advocate of animal rights became a friend, and something of a guru, to George Bernard Shaw, G. K. Chesterton, William Morris and Mahatma Gandhi, so it is true to say that Salt's thinking influenced many of the leading social activists of the last century: those who advocated change primarily for the betterment of humanity.

Furthermore, it was Albert Schweitzer, this century's most celebrated humanitarian, who wrote the essay 'Reverence for Life' that many animal welfarists acknowledge as the cornerstone for the ethic of animal rights.

With such a track record, it is clearly absurd to accuse animal welfarists of being indifferent to human suffering. Indeed, their concern for animals generally reveals an empathy with living beings of whatever species. By contrast, 'humans come first' advocates usually use that argument as an excuse for doing nothing for either man or beast.

4

Scientists versus Saboteurs

Animal rights and the ethics of science

'Animal liberation is also human liberation. We recognize our kinship with all feeling things.' *Henry Spira*

1

The Vivisectors

Konrad Lorenz vivisector test • Morality and practicality of animal experimentation • Criteria for what constitutes legitimate research • High Priests of science.

Author and scientist Konrad Lorenz once proposed a test for scientific researchers. Lorenz suggested that the researcher – in his imagination – should kill and cut up: 1. a lettuce 2. a fly 3. a frog 4. a guinea pig 5. a cat 6. a dog 7. a chimpanzee. At the conclusion of the test, Lorenz wrote: 'To any man who finds it equally easy to chop up a live dog and a live lettuce, I would recommend suicide at his earliest convenience.'

Vivisection and the use of animals in scientific research are the two most volatile areas in the whole debate on the human treatment of animals. This is obviously due to the perceived conflict of interest between human suffering and animal suffering. As one scientific journal once put it: 'It's Fido or you.' A few anti-vivisection groups, in fact, take the stance that, given the choice, it should be you rather than Fido. After all, nobody ever asked Fido if he wanted to volunteer, and the experiments aren't usually designed to do Fido and his family any good.

However, most animal welfare groups point out that very, very few experiments, when looked at closely, could possibly represent a choice between Fido and a human being; a good number of these are for weapons research, to determine a means by which to kill humans more effectively. More importantly, they argue that at least 80% of all experimentation on animal subjects achieves no significant advancement in any area of scientific research.

In 1876, British animal welfare groups first successfully lobbied for a Cruelty to Animals Act. The act was to some degree motivated by those who perceived extreme cruelty in vivisection. At the time, however, only about eight hundred animals a year were used in scientific experiments. A century later, in 1976, more than six million animals a year were being killed in British laboratories. Despite this phenomenal expansion, it took a hundred years to revise the Cruelty to Animals Act – and even then many animal welfare groups felt that it had been so watered down by powerful medical and scientific lobbies that legal restraints were still absolutely minimal – as are policing and enforcement of the Act.

Most animal welfare groups are not arguing for all animal experimentation to cease. They are simply arguing for reasonable restraint; in other words, for an end to painful experiments that are unnecessary, trivial, repetitive and wasteful. They insist that four basic questions should be answered before an experiment is allowed to proceed:

1. What is the purpose of the experiment?
2. Is the experiment properly designed and valid?
3. Is the experiment being duplicated?
4. When an experiment is painful to an animal, is there any consideration of the ratio of suffering to the worth of the experiment?

As a spokesman for the US Humane Society stated: 'If the research is worth doing, the Humane Society does not come out against it. We agree that legitimate research should continue. The argument is about what really constitutes legitimate.'

Organizations like Christine Steven's Washington DC-based Animal Welfare Institute (AWI) feel that there is no possibility of stopping all animal experiments, so they target on legislation that will minimize suffering by guaranteeing reasonable living conditions in laboratories, humane methods of transport, the use of anaesthetics for painful experiments, when possible, and proper veterinary aftercare. Virtually none of these requirements are complied with or enforced. In fact, the almost non-existent protection of animals used for experimentation is dwindling in the face of widespread apathy and the standardization of violent procedures.

In the days of the Roman Empire, conventional wisdom and the expertise of the high priests dictated that the well-being of the human race was dependent on the sacrifice of millions of animals a year. In some temples, as many as three thousand animals a day were killed. When people questioned the considerable cost of the slaughter and argued for a little more restraint, the high priests responded that they alone could be the judges in such specialized matters. Furthermore, recalling the old days of human sacrifice, they ominously and convincingly reminded the people: 'It's Fido or you.'

Scientists and their assistants sacrifice some sixty million animals a year in the United States alone. Less than 15% of these animals are anaesthetized in any way. When, from time to time, America's more caring citizens protest at the expense and the suffering, the scientists fall back on the same ancient arguments as the temple priests.

In Ancient Rome, of course, the day came when the ritual sacrifice of animals was suddenly brought to an end by an imperial decree that converted the Empire to Christianity. Despite dire predictions of forthcoming disaster for failure to sacrifice the beasts, the next morning the sun appeared as usual and the earth remained intact.

One suspects that our own high priests – the scientists – might make a similar discovery if they were to impose a few controls on the more wasteful and ill-considered animal experiments that are performed in the temples of science.

2

Instruments of Torture

Cruelty as a non-issue among scientists • Noble-Collip drum • Other standard battering machines • 'Trauma' and endurance testing • Weapons tests • Psychology of pain.

In 1974, America's PBS television network staged a panel discussion on the issue of primate research between Harvard philosopher Robert Nozick and three scientists. Nozick asked if the fact that hundreds of monkeys may suffer or be killed in an experiment in any way influenced whether or not it was performed. The scientists unanimously agreed that it had no weight whatever. To the question 'Don't animals count at all?' Dr Perachio of the Yerkes Primate Center answered shortly: 'Why should they?' The others concluded that animal experimentation, so far as they were concerned, raised no moral issue at all.

Working within their closed world of scientific research many scientists see no percentage in dealing with the issues of even the most basic of animal rights. In some situations, it even appears that certain human rights are peripheral to the scientific truths being pursued. The degree to which pain and suffering is a non-issue in scientific circles can be roughly gauged by looking at some of the more common devices used in animal experiments. For lack of a better description, these devices have become 'standard instruments of torture'.

One common piece of laboratory equipment is the Noble-Collip Drum, a machine not unlike a tumbledrier. Invented by R. L. Noble and J. B. Collip for shock experiments, it tumbles the unanesthetized animal in a revolving drum in which there are two triangular projections. One carries the animal up the side of the drum allowing it to fall when it reaches the top. It is then picked up by the other projection. To prevent the animal from breaking its fall, its paws are taped together. The drum rotates forty times per minute, with two falls for each rotation, so that in the space of ten minutes the animal endures eight hundred falls. The usual catalogue of injuries includes broken teeth, concussion, bruising of liver, haemorrhaging from mouth and anus. Further examination reveals engorgement of bowels, kidneys, and lungs. Stomach and intestinal ulcers are also common after-effects.

Even though, as early as 1939, a group of British scientists condemned the drum and attempted to have it banned, it continues to be the most popular method of 'traumatizing' laboratory animals.

Beyond the Noble-Collip Drum, however, there is an extensive and varied list of imaginative trauma-inducing instruments. There are for instance restraining chairs and racks for monkeys, cats and dogs, which clamp the animals into immovable positions for months on end, while other even more stressful acts are performed on them. A wide range of mechanical devices are used to batter animals with hammers, or crush them, with clamps. Other machines which inflict electric shock, extreme heat and extreme cold are employed as standard procedures. New instruments are constantly being invented and are advertised to satisfy the research needs of scientists. *Aerospace Medicine* magazine recently advertised a breakthrough in the form of an innovative high-speed trauma-inducing machine capable of 'striking all random surfaces of dogs' hindlegs at a rate of 225 blows per minute'.

Most of these trauma-inducing experiments are supposedly to extend the field of psychological studies. It is argued that their impetus is altruistic and for the general good of the human kind. However, a large and growing body of experimental work is undertaken for the opposite purpose: testing the limits of animal endurance so that humans may be more efficiently tortured or killed. The many experiments on animals, for instance, at Porton Down Trauma Unit in Sussex are conducted almost entirely to help the military learn how to kill humans more effectively.

To this end, Porton Down's Trauma Unit, in common with a multitude of other institutes funded by military research budgets, have had monkeys, dogs, sheep, pigs and steers shot in the brain and body in order to test the damage capability of bullets and explosive devices. After being subjected to lethal doses of radiation, dogs, cats and monkeys have been literally run to death over hours and days in rotary drums charged with electric shock in order to determine how long they can survive. Other animals are tested with biological and chemical weapons as well as more conventional weapons of war, before being subjected to torturous endurance-testing devices. Such massively financed military experiments on animals have perhaps provoked the greatest wrath among animal welfarists

Many trauma-inducing psychological experiments also have at least a suggestion of military use. Take, for example, Dr Roger Ulrich's article based on long-term work at the University of Western Michigan and elsewhere, entitled: 'Pain as a Cause of Aggression'. Dr Ulrich used shock treatment on animals to induce fighting. He found that the higher the ratio of shock in certain animals, the 'more vicious' did the rate of fighting become. In one experiment 15,000 shocks were administered in a seven-and-a-half hour period. In another, shocks were delivered at a relatively lower rate over eighty days. He also used extreme heat, cold, intense noise, castration, cut whiskers and removal of the eyes to provoke aggression.

As to the direction in which this research may be moving, Ulrich has already admitted that he looks forward to finding some means of studying humans: 'Naturally the moral and practical difficulties are tremendous. Yet, as our knowledge of aggression in lower animals progresses and as more and more feasible methods of studying aggression in humans are developed, a clear picture should emerge.'

3

Edison's Elephant and Others

Thomas Edison fries an elephant • Baboons and pigs in car crashes • Polar bears in oil • Classroom indoctrination in cruelty.

In an early newsreel film, the famous inventor Thomas Edison proposed a unique 'experiment'. A full-grown African elephant was led by its trainer and made to stand on a large iron grid. Edison's technicians then switched a pair of large circuits which released a massive electrical charge through the grid and the elephant. The screaming beast toppled slowly forward as its feet smoked and its flesh became welded to the iron. The writhing trunk twisted and jerked as the huge animal literally fried on the grid. The film of this event resembled nothing so much as the spectacular destruction of the *Hindenburg* zeppelin collapsing in flames.

Admittedly, some bizarre experiments have proved useful in the advancement of scientific research, but this was no experiment at all. It was a spectacle: a deadly circus act that proved nothing at all. Edison's 'spectacle' was just one in a long tradition of cruel and pointless experiments. Only occasionally is the public made aware of such projects – most of which are dressed up in more formally scientific specifications – despite the fact that they are nearly always publicly funded.

A couple of recent examples should reinforce this point. From 1971 to 1980, the French government pumped over $5 million into a laboratory complex in order to conduct experiments on baboons, macaque monkeys and pigs. In these experiments, designed to study automobile crash impact, 31 baboons and an unknown number of pigs were killed in 40-m.p.h. sledge crashes. A larger number of macaque monkeys were also killed when their skulls were bludgeoned repeatedly with a hammering device that carefully measured fatal impact levels.

In broad terms, the experiments established that crashing a car into a wall at 40 m.p.h. is an unhealthy activity especially for a pig or a monkey. But as the anatomy of these animals is sufficiently different from humans for the injuries sustained to be nothing like those in a human car crash, it is difficult to see what else the experiment might be expected to prove. And as few pigs or monkeys drive motor vehicles, the research is unlikely to benefit their species.

Perhaps the only positive outcome was that other crash experimentation laboratories did not leap to the defence of these tests. Many critics made the point that there was hardly a shortage of human car crash victims on which to gather relevant data. Others pointed out that mechanical dummies in human shape have long been used in such experiments, and are far more accurate in determining impact and injury than any animal subject. That may be so, said one spokesman in a rather offensive rejoinder, but the animals were 'cheaper than using mechanical simulators'.

Another peculiar experiment was one funded by the Canadian federal government in 1980 and conducted in Manitoba. In this experiment, which cost in excess of $120,000, three captive polar bears were forced to swim through a tank filled with crude oil and water to determine the effect of oil pollution on white-furred animals. To no one's surprise, the oil entirely coated the animals' fur. The bears then licked the oil from their coats and ingested sufficient petroleum to induce kidney failure and death. The conclusion of the experiment was: 'Keep polar bears away from oil-slick areas.' To add insult to injury, the researchers went on to 'urge further studies', despite the public uproar over the pointless deaths of these supposedly protected animals.

Less than one month later, undeterred by public antipathy, a Canadian university team was being financed through an American grant of a million dollars to do a similar study on dolphins. It was headed by Dr Joe Geraci of Guelph University who had just spent five years in the employ of the Canadian government immersing ringed seals in crude oil. As a result of his experiments over a hundred seals died or had to be destroyed.

What particularly enraged environmentalists about these studies was the fact that a huge amount of information on marine life and oil pollution has already been gained from the hundreds of massive oil spills of the last few decades. Furthermore, oil spills happen with such regularity that Dr Geraci and his colleagues would have had to wait only a short time for another to occur, and experiments could be conducted on site.

The reason why such experiments are permitted, and why so few questions are asked to justify them, is that all animal researchers have been consciously trained to view their animal subjects as 'tools for science'. Laboratory students are required to do repetitive and boring experiments which kill millions of animals a year, not to teach them anything new, but rather to indoctrinate them into a problem-solving process. There is seldom an effort to view the problem in a way that might make the experiment redundant, and rarely is thought given to ways in which the animal's suffering might be reduced once the experiment is embarked upon.

4

Harlow's Hell for Monkeys

Harlow's psychotic monkey experiments: maternal deprivation, 'monster mothers', homicidal mothers • 'Well of despair', 'tunnel of terror' • 'Psychological death' • Evaluation of experiments

Most researchers claim to have no animosity towards animals, while some actually claim to be very fond of them. Some, however, are very blunt in their attitude. Dr Harry Harlow, an experimental psychologist from the University of Wisconsin, is one such plain speaker. At a 1974 conference in Pittsburg, Dr Harlow was quoted as saying: 'The only thing I care about is whether the monkey will turn out a property I can publish. I don't have any love for them. Never have. I really don't like animals. I despise cats. I hate dogs. How could you like monkeys?'

This attitude might be harmless enough in an ordinary citizen with no animals in his care, but Dr Harlow has spent over three decades experimenting on monkeys. He is well known for his experiments in inducing psycho-pathological behaviour in monkeys. Within the scientific community, Dr Harlow has received nothing but praise for his work.

Dr Harlow's best-known experiments in psychotic monkeys involve the infant-mother relationship. In 1965, Dr Harlow worked with a team which for ten years studied the effects of social isolation on infant monkeys. Some were raised from birth in bare wire cages and subjected to maternal deprivation. Others were raised in stainless steel chambers, without any animal or human contact. Predictably, both groups of animals were reduced to cowering little heaps whose primary response to any social contact was fear.

Dr Harlow, believing mere deprivation insufficient, attempted a series of more complex experiments involving surrogate mothers. These allowed deprived infant monkeys to become fixated on cloth dummy monkeys as mother substitutes who eventually turn into monsters. Dr Peter Singer, in his book *Animal Liberation*, examines Dr Harlow's activities in some detail, and quotes him on his surrogate mother experiments:

> The first of these monsters was a cloth monkey mother who, upon schedule or demand, would eject high pressure compressed air. It would blow the animal's skin practically off its body. What did the baby monkey do? It simply clung tighter and tighter to the mother at all costs. We did not achieve any psychopathology.
>
> However, we did not give up. We built another surrogate monster mother that would rock so violently that baby's head and teeth would rattle. All the baby did was cling tighter and tighter to the surrogate. The third monster we built had an embedded wire frame within its body which would spring forward and eject the infant from its ventral surface. The infant would subsequently pick itself off the floor, wait for the frame to return into the cloth body, and then cling again to the surrogate. Finally, we built our porcupine mother. On command, this mother would eject sharp brass spikes over all the ventral surface of its body. Although the infants were distressed by these point rebuffs, they simply waited until the spikes receded and then returned and clung to the mother.

None of these results., Dr Harlow concluded, proved anything surprising, except the ability of infant monkeys to take extreme punishment. As with humans, injured monkey infants have no other source of comfort than their mother, even if that mother causes them pain. A study of battered children would tell a psychologist the same thing and would prove far more useful to human psychology than Harlow's monkey experiments. However, Harlow and his fellow researchers persisted. One set of experiments used isolated and deprived real monkeys instead of dummy monkeys as monster mothers. Because of their deprivation, these surrogate mothers had no idea of their roles. They either ignored the infants or became insanely hostile, and finally homicidal.

Dr Harlow set up the experiments, then watched and took notes: 'One of their favourite tricks was to crush the infant's skull with their teeth.' Another 'trick' was 'that of smashing the infant's face to the floor, then rubbing it back and forth.'

From there, Dr Harlow went on to create such evocatively named experiments as 'the well of despair' and 'the tunnel of terror' for his monkeys. In these, Harlow successfully induced 'severe and persistent psychopathological behaviour' through extreme despair and terror.

Finally, Dr Harlow introduced experiments later widely copied and expanded on which induced 'psychological death' in monkeys. In one such group of experiments, Dr Harlow's team provided rhesus infant monkeys with surrogate mothers made of towelling which could rapidly be chilled to freezing point to stimulate rejection through deathly cold.

At the end of the day, how valuable did these experiments prove to be? Certainly most social workers agree that Harlow's research money would have been better spent on work with already deprived children. Perhaps Dr Harlow himself should have the final word on this. For twelve years he was editor of the *Journal of Comparative and Psychological Psychology* which established some near-records in the number of painful experiments published in its pages. At the end of his tenure, Dr Harlow wrote a farewell statement in the magazine. In it, he said that he had reviewed some 2,500 manuscripts during his editorship, and from these he had concluded that 'most experiments are not worth doing and the data attained are not worth publishing'.

5

Animal Liberation Front

Animal rights protestor's suicide • Animal Rights Militia letter and fire bomb attacks in Britain • ALF cells in Canada, US, France, Holland, Germany • Extremists *vs* moderates • Police infiltration and the jailing of 'General Lee' and others.

On a mild and pleasant day in March 1986, a young man sat in a quiet urban public park in Colchester. He had arrived at a state of absolute despair over the issue of animal experimentation in Britain, and decided to stage a personal protest of the most extreme kind. He doused himself in petrol and set himself alight. He soon died in the fiery protest.

This protest, commentators argued, was a futile act by an unbalanced and disturbed individual. However, it must also be said that it was indicative of the extremity of human emotions on the issue of cruelty in scientific laboratories. Anti-vivisection groups in recent years in most Western nations have markedly increased their membership. However, the most dramatic and obvious indication of deepening concern has been the appearance of numerous animal rights saboteur organizations who specifically target scientific institutions and those who support them.

The most extreme of these is probably (ARM), the Animal Rights Militia, which in 1982 and 1983 sent letter bombs to Britain's five national party leaders and half-a-dozen prominent scientists known to be vivisectors. One exploded in the Prime Minister's office, slightly injuring the office manager, and another exploded in a surgeon's home.

ARM is the shadowy anonymous terrorist cell faction of ALF, the Animal Liberation Front. The membership of both organizations is nominally secret, but ALF has its public face. By 1980, two ALF executives Ronnie Lee and Cliff Goodman had served four-year sentences for numerous actions of sabotage, arson and theft against vivisectors and animal laboratories. Because he was known to the police, Ronnie Lee ceased to be an operative in the organization, but essentially became its press officer, making public statements after ALF or ARM had struck. At least ten other ALF and ARM members in Britain had been jailed by 1985 for their raids on animal experimenters' laboratories and private homes throughout the country. These actions amounted to several million pounds' worth of damage, and were the cause of considerable alarm in the scientific community.

By 1986, ALF had extended its mandate beyond animal researchers and attacked any businesses involved in broadly defined 'animal abuse'. They had now entered a 'vengeful vegetarian' phase in which anyone involved in the meat trade, from butchers to retailers, also became legitimate targets. An extensive fire bomb campaign hit major targets in Cardiff, Birmingham, Sheffield, Liverpool, Croydon and London. By 1986, ALF and ARM had drawn too much attention to themselves for their own good. Scotland Yard's C-13 Anti-Terrorist Squad and the C-11 undercover team worked in co-operation with local police forces to step up surveillance on known ALF activists.

In 1986 the Anti-Terrorist Squad, using extensive electronic surveillance, uncovered the main ALF fire-bomb factory in Sheffield and managed to arrest several major figures including according to police their 'master bomb-maker', Ian Oxley. The ALF trial in Sheffield in 1987 resulted in the conviction of six men and two women. Ian Oxley was sentenced to four years' imprisonment as was ALF's northern 'commander', Roger Yates – although Yates escaped and went into hiding just before the end of the trial, so was sentenced in absentia. The most severe sentence went to Ronnie Lee, who was proved by the undercover team to be not just the press officer but the active director of the whole organization. He had been nick-named 'The General' by the police, and was sentenced to a full ten years' imprisonment.

Although the Sheffield trial must certainly have dealt ALF a severe blow, there is no sign that its other members are going to give up their activities and there were a number of reprisal raids since the trial. Then, on 20 December 1988, ALF targeted the fur trade and in a single night set off fire bombs in five cities across England. The most devastating of these attacks entirely burned down a multi-million pound department store in Plymouth that was owned by House of Fraser and Harrods – both of which also suffered fire bombing that night.

Meanwhile ALF, which in its first decade of operation, claims to have carried out 10,000 actions, has spread to several other countries. There is a Dutch ALF called the Dierenbevrijoings Front, and the French have at least two equivalent organizations: Katpat and SNDA, the Société National pour la Défense des Animaux. There have been a large number of laboratory raids by ALF in Canada, and the American ALF has claimed responsibility for the burning down of a five-million-dollar research laboratory complex belonging to the University of California.

It is difficult to deny that extremist actions by animal rights activists serve to publicize issues such as animal experimentation that are generally ignored by politicians and the media alike. However, animal welfarists take two conflicting perspectives on such actions. One side argues that extreme actions, especially those that have the potential to harm people or even their property, cause extreme public antipathy for the whole animal welfare movement. They believe that politicians take advantage of extremist actions by using them to condemn all animal welfarists as cranks who value animals above humans. They also believe that the police will use such actions as an excuse to infiltrate all animal welfare organizations and to oppose even non-violent protest actions.

In 1989, ALF moved up into the terrorist 'big league' in Britain, when it appears to have used sophisticated plastic explosives to blow up a Bristol university building.

ALF activists and other extremist organizations on the other hand, maintain that legal and non-violent actions have not achieved any significant improvement: after a century of campaigning, ever more animals suffer and die each year. Even if extremist actions fail to win over public opinion, they argue, the issues will become broadly known and debated. At least then, they believe some change may occur as a compromise by politicians to more moderate demands.

ALF and similar organizations have become so legally hot that virtually all their meetings, if given any public notice, are under constant surveillance. In fact, surveillance was already considerable in the early 1980s. At that time, one investigator attended a cell meeting in London and, later, he methodically tracked down each of the new recruits for further investigation. One was traced back to metropolitan police headquarters, another to the *Daily Mail*, a third to Scotland Yard, a fourth to a private investigation agency. Apart from the organizer who called the meetings, the investigator found that the entire gathering was made up of 'bent' informers.

6

The Poisoners

Target: 'Lethal Dose-50' as standard toxicity test • Kills millions of animals each year • Henry Spira and the Coalition to Abolish LD-50 • Thalidomide and the failure of LD-50 • Bad science and bad law.

As has been seen, most campaigners for animal rights tend to reject the practices and aims of such organizations as the Animal Liberation Front. They realize the futility, even in the limited area of scientific research, of attempting to change a whole system overnight. They know they must concentrate their efforts on a few obviously vulnerable target areas in order to achieve a modicum of change. One of these is toxicity testing.

In 1982, a major conference on animal research was held in Britain. It was a gathering of scientists, researchers and academics – professionals in the field of animal research – some working for such organizations as Shell and Huntington. The eventual outcome of the conference was a report on the state of animal research.

The report, three years in the making, was undertaken for an organization called FRAME, the Fund for the Replacement of Animals in Medical Experiments. The conclusion it reached was that current methods of toxicity testing are 'bad science and bad legislation'. The scientists agreed that the number of tests, which result in four million animal deaths a year in Britain alone, could be immediately cut in half without discernible effect by extremely minimal changes in practice and legislation. In fact, after three years' deliberating, it was concluded that a better and more scientifically logical approach would result from a dramatic reduction of experiments.

The main target was the LD-50 test, described as a 'mindless insistence on crude body counts'. According to one conference delegate 'Governments like LD-50 because it is a number like a traffic speed limit. LD-50 stands for 'Lethal Dosage – 50%' and is a standard test for toxicity of consumer products which has been in use for fifty-five years. The idea is to force-feed a substance to a test group of animals until 50% of them die. The results of the experiment produce a number that fits into a rating system.

For instance, a household product is considered 'highly toxic' if it produces 'death within 14 days in half or more than half of a group of test animals', according to the US Food and Drug Administration. These tests are most often conducted on mice, rats, guinea pigs, rabbits, dogs and monkeys.

For some substances like cosmetics, which are not particularly toxic, the LD-50 test is clearly absurd. There is little chance of any human swallowing enough face powder or talcum power to poison themselves, yet such products have to be force-fed to dogs and rats by stomach tube in massive quantities for prolonged periods. Eventually, the internal organs are blocked or ruptured; even though such blockages in fact have nothing to do with the toxicity of products, a painful lingering death is effected in the case of 50% of the animals and a number is recorded for statistical consistency. The remaining 50% are then mercifully put to death.

The belief of the majority of animal rights groups is that LD-50 is wasteful and absurd in its blanket application. Many have joined an organization called the International Coalition to Abolish LD-50. One of the Coalition's leading campaigners and activists, Henry Spira, argues: 'We are aiming at better protection of the public. The LD-50 is not just cruel, it is also an archaic method of testing toxicity. The question the test puts is: 'How many bars of pure ivory soap does it take to kill half a batch of 20, 30 or 60 dogs?' This is ridiculous.'

The Coalition's immediate aim is to lobby and campaign for the US chemical industry (which extensively uses the LD-50 test) to subsidize a $10 million research fund to establish an alternative testing method. Spira's argument is that the industry, with earnings of $450 billion per year, has a moral obligation to use some of its profits to reduce the suffering of the animals sacrificed to facilitate the marketing of its products. Furthermore, he believes that these producers are likely to bow to such a campaign because they are companies whose sales depend upon good public relations.

Virtually all household products – including bubble bath, nail polish, oven cleaner, deodorant, hair spray, antifreeze and birthday candles end up being tested on animals to sometimes horrific effect. Innocuous products often result in terrible suffering. One nasal decongestant, for example, was tested in such a way that test animals only died after a prolonged cycle which included salivation, convulsions, bleeding from the nose and mouth, diarrhoea, loss of muscle control and prolonged vomiting. The test animals took weeks to die and included 96 rabbits, 24 monkeys, 5 cats, 376 rats and an unreported number of mice and dogs.

However, whenever the subject of LD-50 is raised, the argument for human welfare is used to counter concern for animals. Sensational examples like the Thalidomide babies are brought up to discourage criticism of the practice and to demonstrate the absolute necessity of LD-50 to protect the public from drugs put on the market.

But the fact remains that Thalidomide *was* tested on animals before it was marketed. The toxicity tests were extensive and it was found to be 'an almost uniquely safe compound'. Even after the drug was known to cause birth defects in humans, further toxicity tests on pregnant dogs, rats, monkeys, cats and virtually every other type of animal produced no irregularities. Only one special breed of rabbit produced deformities.

This is exactly the point about LD-50 testing which the Coalition is arguing. Insulin produces deformities in mice and rabbits but not in humans. Thalidomide is harmless to most animals but has proven disastrous when ingested by humans. Penicillin is toxic to rabbits and guinea pigs but not to humans, although some are allergic to it. Species are very different from one another and for this reason, LD-50 is unscientific and unreliable as a measurement of safety or toxicity. New methods must be developed.

What should these methods be? That is the question the Coalition wants answered. There are many options, but a single standardized system that applies to all products needs time and money for development. The Coalition suggests that a standardized system of tissue culture tests and computer models of biological systems could, and should, be developed for reasons both human and humane. Meanwhile, the immediate implementation of a stricter checking system to eliminate duplication of experiments would be one of several much-needed reforms in current testing procedures. Other measures, such as more extensive use of anaesthetics on animal subjects, would also help to greatly reduce the amount of unnecessary suffering.

7

The Blind Rabbit Campaign

Draize rabbit test for cosmetics • Targeting: 'How many rabbits does Revlon blind for beauty's sake?' ad • Alternatives and a limited victory • Beauty Without Cruelty • The Body Shop.

'Animal liberation is also human liberation.' This is the credo of animal rights activist Henry Spira. However, Spira's idealism is offset by hard-headed street-smart campaign knowledge gained in the civil rights movement of the 1960s and the animal rights movement of the 1970s. In the struggle, he points out, 'the fundamental lesson is that the meek don't make it. Audacity must be fused with meticulous attention to programme strategy. Effective actions are the result of people gaining confidence in their capacity to effect change. Our confidence comes from winning limited victories which in turn become the stepping stones for further struggles and greater victories.'

Before Spira helped create the International Coalition to Abolish LD-50, he focused on a much smaller and more vulnerable aspect of animal experimentation in order to create public awareness of the issue. This was the American Museum of Natural History's cat experiments. For twenty years the museum had carried out ludicrous and cruel experiments on cats. Taxpayers' money was being used to blind, deafen and mutilate the brains and reproduction organs of cats and kittens in order to study their sexual performance. Spira's campaign rapidly drew public attention to the issue, and finally forced an end to two decades of painful experimentation.

From there Spira moved on to a much larger target before initiating the LD-50 Coalition. This was the Draize Eye Test which had been continually in use for forty years. Spira believed it was a good middle-ground target which could be easily recognized by the public as unnecessary and cruel; above all, the companies using the test, which included Revlon, were vulnerable to unfavourable publicity.

So began the 'Revlon Blinds Rabbits' campaign, with headlines like 'How Many Rabbits Does Revlon Blind For Beauty's Sake?' This was followed up by a picture of a disorientated rabbit with eye-patches, and the rhetorical question: 'What is the Draize Rabbit Test?'

> Imagine someone placing your head in a stock. As you stare helplessly ahead, unable to defend yourself, your head is pulled back. Your lower eyelid is pulled away from your eyeball. Then chemicals are poured into the eye. There is pain. You scream and writhe helplessly. There is no escape. This is the Draize Test, the test which measures the harmfulness of chemicals by the damage inflicted on the unprotected eyes of conscious rabbits, the test that Revlon and other cosmetics firms force on thousands of rabbits to test their products.
>
> There are no ifs or buts and maybes. These bunny rabbits are bred to be blinded. It's happening today. In 1978, 2,692 rabbits were forced into Revlon's research center in the Bronx. And none was given any pain relief even though the Draize manual suggests that eye mutilations be observed for twenty-one days. Their only salvation from eyes slowly eaten away by chemicals is death.

The reason for the choice of rabbits for the Draize Test is because rabbits can't 'cry'; that is, their tear ducts produce very little fluid to work out impurities. Consequently, if any irritant is placed in the eye, and the animal is restrained, the substance will remain there. Furthermore, a rabbit eye is far more sensitive than a human eye, and therefore ideal for maximum possible damage.

Because of these properties, rabbits are used to test cosmetics, pesticides, detergents and such substances as oven cleaner. For up to seventy-two hours, the rabbits are put in stocks and subjected to increasing amounts of irritants. They then remain for observation for at least twenty-one more days. Positive reactions are: 'ulceration of the cornea, inflammation of the iris, haemhorrage, gross destruction of the conjunctivae, obvious swelling with partial eversion of the lids, a diffuse crimson-red with individual vessels easily discernible'. Essentially, the eye is eaten away by the chemical. After this prolonged period of painful mutilation without anaesthetic, the rabbit is put to death.

Statements by scientists seem to be contradictory: 'No unnecessary pain or distress is inflicted upon the subject animals . . . therefore no anaesthetic, analgaesic or tranquillizing drugs are used'. If pain is judged 'necessary, then pain relief is unnecessary'.

Why is no anaesthetic or tranquilliser used? It is argued that no painkiller can be given to animals undergoing the Draize Test because 'it would interfere with the interpretation of results'. Dr Dallas Pratt, MD, in his investigation of the Draize Test in *Alternatives to Pain in Experiments on Animals*, has pointed out: 'All the analgaesic would do is dull the sensation of pain; surely the results would still be plain to see on the blistering skin and in the haemorrhaging eye?'

A further argument against the Draize Test, and one which is levelled at it by many within industry itself, is that the results are irregular. A medical research unit of Esso found 'extreme variation' in the way labs evaluated rabbit reaction to standard irritants and industry investigators concluded that these tests 'should not be recommended as standard procedures in any new regulations. Without careful re-education these tests could have unreliable results '.

In the 'Revlon Blinds Rabbits' campaign, Henry Spira welded together a coalition of four hundred animal welfare groups. This coalition did not demand that Revlon cease marketing cosmetics tested in painful experiments but settled on an achievable goal: Revlon should pay for the development of new methods of cosmetic testing – particularly tissue cultures – which could replace the Draize Test. By December 1980, Revlon had capitulated and volunteered to contribute $750,000 to a research fund. Avon followed with another $750,000 and Estee Lauder contributed $250,000.

Spira had achieved his 'limited victory' and a 'stepping stone for further struggles and future victories'. As he said on the announcement of the Revlon donation: 'Even though it might be a long road, at least we've started on it.'

This might be an appropriate summation of the work to date of many groups in the animal rights movement, one of which, called Beauty Without Cruelty, has gone to the lengths of creating a line of cosmetics which are neither made of animal products nor tested by animal experimentation methods. Much to the surprise of many cosmetic firms, the marketing of such humane and ecologically sound products need not be commercially disastrous. Probably the most admirable example of this has been the phenomenal success enjoyed by Anita Roddick's *Body Shop* chain, exactly because it has clearly declared such a humane and ecologically sound philosophy.

8

Maryland Monkey Trials

Dr Taub's monkey trial conviction • Alex Pacheco's undercover evidence • Police seizures • Crippling of monkeys • Questionable research value • Darwin's view.

On 23 November 1981, animal welfare agencies announced a 'landmark victory in the struggle for animal rights' when the trial of psychologist Dr Edward Taub ended with his conviction on six counts of cruelty and a fine of $3,000. For the first time, a scientist had been tried and convicted in an open court on the issue of cruelty within the laboratory.

This trial was made possible by the undercover work of a twenty-three-year-old activist named Alexander Pacheco, who volunteered to work as an assistant to Taub over a four-month-period. During that time, Pacheco gathered a four-minute colour film, seventy photos and a great deal of incriminating documentation to prove that the monkeys in the lab suffered all manner of injury: broken bones, exposed muscle, chronic infection and exposed root canals. These injuries received virtually no treatment, and he also charged that the laboratory assistants were encouraged to 'torment and frustrate' the animals in the testing process. Based on Pacheco's documentation, an unprecedented police raid took place and Taub's seventeen monkeys were removed from his Institute of Behavioral Studies in Silver Springs, Maryland.

Initially, Taub was charged with seventeen counts of cruelty: one for each monkey. However, the judge discounted obvious physical damage and suffering as subjective and inadmissible. What *was* found to be admissible was that six monkeys had severe medical problems which, under the letter of the law, required treatment by a vet. Taub had kept no medical records and, by his own admission, no vet had visited the lab for at least two years.

Even the highly conservative national Institute of Health, which rarely censures its members, eventually judged Taub's laboratories 'grossly unsanitary', and concluded that he had failed to provide proper medical care. The NIHS's conclusion resulted in an almost unprecedented suspension of financing on the grounds that its guidelines for animal care had been violated. This measure alone carried far more weight than any court conviction, for the grant suspension meant that Taub lost $115,000 in government funding.

There was no legal requirement to prove the validity of the tests themselves, but from the facts that emerged at the trial animal activists were dismayed by the pointlessness of the monkeys' suffering. Dr Taub had, over eleven years, been conducting 'differentiation research': the modification of body parts for performance of certain functions. In the process, he had spent hundreds of thousands of dollars of public money for experimentation in which scores of monkeys were crippled or killed.

For the purposes of the experiment, nerves controlling the animals' limbs were severed and undamaged comparable parts restrained to see if this encouraged rehabilitation of the injured limb. The stated aim of this research was that it is intended to add to our knowledge in order to benefit human stroke victims. However, this kind of therapy has been extensively tested for over a century on human victims of cerebral injury so the purpose served by subjecting non-humans to the same procedure is beyond the comprehension of most researchers.

After a lengthy appeal trial and much deliberation by the jury, which was out for twenty hours, Taub managed to avoid all but one charge of cruelty to a monkey named Nero. Yet, even so, the scientist was unrepentant and declared himself – rather than his victims – to be a martyr of science. Deflecting attention away from his research, Taub focused instead on the necessity for free inquiry in the name of scientific advancement. In a rather hyperbolic association, he likened his position to that of Galileo, who had been persecuted by the Catholic Church for his view of the universe.

Galileo's views on monkeys and vivisection are unknown because his own research was both humane and cerebral. However, we do have the opinion of one other great scientist with revolutionary ideas who was well acquainted with both monkeys and the procedures of vivisection. Charles Darwin was extremely frank in his views on this matter. 'You ask my opinion about vivisection. I quite agree that it is justifiable for real investigations on physiology, but not for mere damnable and detestable curiosity. It is a subject which makes me sick with horror.'

9

More Monkey Raids

Britain's monkey trial • Raid on Royal College of Surgeons lab • Successful cruelty prosecution.

Britain recently had its own monkey show trial. Though less dramatic than the American version, its conclusion is uncomfortably sobering. In February 1985, a test case was brought against the Royal College of Surgeons (RCS) by the British Union for the Abolition of Vivisection (BUAV), under the 1911 Protection of Animals Act. BUAV charged that the RCS was responsible for inflicting unnecessary suffering on monkeys at its research farm in Kent. This legal action was, of course, only made possible because animal activists broke into the laboratory and stole accident-report files made out by the institute's own technicians, which they passed to BUAV.

BUAV was therefore able to prove that the living conditions of animals in their twenty-year-old cages were inadequate. Some fifty-two monkeys had trapped their limbs in cages, causing severe damage. A large number had suffered dehydration because of conditions of extreme heat and inadequate water supply. Two monkeys left without water died after breaking their arms while trying to reach water dishes outside their cages. As the basis for the action was, in fact, records carefully prepared by the staff of the Royal College of Surgeons itself, it was an obvious case of self-incrimination. The court had no option but to find the RCS guilty of the charge of 'inflicting unneccessary suffering' and a nominal fine of £250.00 was imposed.

BUAV was understandably happy about the results of its test case, even though the fine was not in any way punishing, and vowed to act swiftly on any further information it received. However, along with most other animal welfare groups, BUAV was forced to conclude that the case demonstrated that widespread cruelty was routinely and knowingly practised by even the most prestigious of scientific bodies, and that internal policing and controls were a charade to lull the public into inaction.

The response of the Royal College of Surgeons was typical of the medical profession and the scientific research establishment on such issues. The RCS ventured to comment that the prosecution had resulted in at least one positive action on its part: the allocation of £20,000 in funds for repairs and improvements at its research farm in Kent. Would these changes improve the conditions for the animals in its keeping? No, it would not alter their conditions in any appreciable way. The money was to be used primarily for improvements to the security system at the farm. Furthermore, the RCS warned, they would spend even more on security if animal liberation groups continued to infiltrate their research laboratories.

The Royal College of Surgeons is part of a privileged scientific establishment which appears to consider itself as beyond criticism, and is contemptuous of even the most basic ideas of 'animal rights'. It is not, however, an issue that will go away.

10

Human Experiments

Nazi scientists • Japan's Unit 731 • Morality and ethics of science.

Any discussion of the issues of animal rights and laboratory ethics must eventually come to terms with the uncomfortable issue of human experimentation. One only needs to look at the record of the scientific and medical community in any situation where taboos against human experiments have been relaxed or removed to see how unsuited it is to policing itself.

Nazi Germany is, of course, the most obvious example: hundreds of doctors and scientists used hundreds of thousands of captive men, women and children to set up and conduct some of the most horrific and sadistic human experiments in history. Similarly, in war-time Japan, hundreds of scientists from the best medical colleges and universities flocked to the infamous Unit 731, the germ warfare laboratories in Pingfan, Manchuria, in order to take advantage of the opportunity of conducting painful and fatal experiments on thousands of Asian and European human subjects.

After the war, 23 German SS doctors and scientists were sentenced to death and many others imprisoned. (Unfortunately, the worst of these, Dr Josef Mengele – the 'Angel of Death' at Auschwitz who was responsible for the torture and death of 400,000 prisoners – was among the many who escaped retribution.) It must be remembered, however, it was not the scientific community which condemned these men. Indeed, in Japan, none of the war criminals of Unit 731 was prosecuted. Many are still alive and are prosperous and respected pillars of the scientific and medical profession.

The terrible truth is that many of these brutal human experiments were an open secret. Without even a mild protest, thousands within the medical and scientific community calmly read and listened to lectures on experiments that were clearly nothing less than sadistic murders.

The issues of animal rights in laboratories are all a part of a larger question about ethics and morality in the pursuit of knowledge. The fact is, there is simply no inherent morality in the pursuit of scientific knowledge. The excesses of the Nazi scientists alone should make us absolutely certain that it must be the responsibility of society as a whole – not the scientific community on its own – to provide the moral guidelines for research.

5

The Communication Corps

Interspecies communication and its possibilities

'Beasts abstract not' *John Locke*

'Gorilla damn fine animal' *Koko, the gorilla*

1

The Talking Beast

Washoe, the first 'talking' chimp • Use of Ameslan
sign-language of the deaf • 160-word vocabulary
• Conversations and concepts.

In 1970, an animal straight out of myth and legend was casually sitting
on a bench at the University of Nevada, munching a banana, as
scientists and trained observers came from all over America to have a
closer look. Scientists were not the only people interested; overnight
the banana-munching beast had become an international celebrity.

The reason for the fame of the beast was the fact that she was the
first animal to 'talk' to humans in a way that could be understood as
language. The chimpanzee's name was Washoe and she was the star
student of Beatrice and Allen Gardner, a husband-and-wife team of
psychologists who, since 1966, had been teaching the chimp Ameslan
or ASL (American Sign Language), the language of the deaf and dumb.
Through the medium of this language, history saw humans and beasts
having their first conversations since the age of fables.

Until the Gardners began their experiment in 1966, the only
attempts to give apes 'language' had been very typically human-
orientated. Most experimenters had assumed that if humans could
learn to use their voice in a controlled way to shape words and
construct language, so, too, could the ape, unless it was just too stupid.
Consequently, when even the brightest of these chimpanzees brought
up in human homes managed to articulate just four words: 'momma',
'poppa', 'cup' and 'up', chimp experimenters concluded that apes were
just not capable of using language as such but could merely parrot a
few words on cue. Furthermore, if that was all one was after, one
might just as well attempt a conversation with a parrot, a toucan or a
budgie all of whom have vast vocabularies compared with these
chimps, while having very little in the way of a brain. Indeed,
comparison between a chimp and a budgerigar on this score might
lead one to conclude that the less brain the better, so far as
conversation goes.

The Gardners were the first people to spot the extraordinarily simple reason why chimpanzees were unable to talk back to us. It was, with hindsight, obvious: the chimpanzee voice box (pharynx and larynx) is simply not suited to speech. Unlike many birds who have a wide range of voice capabilities and mimicking powers, chimps were limited by anatomy rather than intelligence. The Gardners recognized at the same time that apes had considerable dexterity and expressiveness in hand gestures and it occurred to them that ASL might be a less limiting means of apes acquiring language.

The success of their experiment, and the ability of Washoe to acquire the Ameslan language, astonished everyone, including the Gardners. After four years, Washoe had acquired a working vocabulary of 160 'words' and was carrying on two-way conversations with humans.

Beyond learning the names of things Washoe was able to 'read' questions, construct sentences and make demands. At first these were limited to the 'Gimme food' or 'Gimme drink' kind but the statements became increasingly complex. One day Washoe was given her mug which unexpectedly had a small doll in it. She reacted by signing: 'Baby in my drink'.

Later in the experiment, she would take her humans to the refrigerator and make the signs: 'Open food drink', and addressing her trainer, Dr Roger Fouts, she often requested: 'Roger Washoe tickle'. She also showed an ability to create new word combinations, such as calling a duck a 'water bird'. It often seemed to her trainer that Washoe made chimp jokes of sorts, which she usually confirmed by signing: 'Funny'. Then she started to use 'toilet words' as a means of swearing. The sign for faeces and urine being 'dirty', when Washoe became annoyed at a rhesus monkey she often signed 'dirty monkey'.

Another of the chimpanzees trained in Ameslan by Roger Fouts was clearly able to discern the opposite meaning that the sentence structure and word orders implied in such constructs as 'Roger tickle Lucy', and 'Lucy tickle Roger'. Like Washoe, Lucy did creative cursing and invented word constructs; she described watermelon as 'drink fruit' and, after experiencing the burning taste of a radish, she always called them 'cry hurt food'.

2

The Computer Chimp

Lana, first chimp to learn computer language called Yerkish • Lana's logic and grammar lessons • Lana's loneliness.

Although certainly the most successful language for apes, Ameslan was not the only one that chimps were to master. Soon there was 'Yerkish'.

This came about because some scientists were concerned that in teaching Ameslan to apes, there was an element of cueing involved. That is, the apes were anticipating through facial clues or body language what was expected of them. To remove this element as a possibility, Duane and Sue Rumbaugh, with others at the Yerkes Regional Primate Research Center in Atlanta, developed a sturdy computer console programmed in a computer language called Yerkish.

In front of this large computer panel with its numerous keys in nine geometric shapes and seven colours, they placed a chimpanzee called Lana. Lana was taught sequence or grammatical structure in requesting things. A sentence would typically be typed into the computer: 'Please-machine-give-piece-of-banana', or 'Please-Beverley-move-ball-into-room', or 'Please-Tim-groom-Lana'.

More than one experiment resulted in unexpected results. On one occasion, her trainer, Timothy Gill, decided to see what would happen if he mixed up her sentences. So when Lana typed in her sentence Gill fed in a different word from his control panel. When the machine did not respond as expected, Lana became annoyed; she looked round and, seeing Gill at the other control panel, promptly typed in: 'Please-Tim-leave-room'.

From the point of view of behavioural scientists, it seems that Yerkish is more immediately acceptable than the 'subjective' Ameslan. It also seems that it can be quickly learned. As its inventors noted: 'After a mere six months . . . Lana could complete correct sentence beginnings and cancel ungrammatical ones. Since then, Lana has demonstrated that she can do considerably more.'

Indeed she can, but from the chimpanzee's point of view, Yerkish computer language is a sterile way of learning. No human child ever learned language in such a way and it is unlikely that the average child could, or would, have the motivation to learn very much like this. The motivation for learning is not so much to do with mental ability as emotional reward.

Perhaps Lana herself has best demonstrated the restrictions of sterile, mechanical teaching. When left alone in her room late one night, with only the computer for company, she plaintively typed in: 'Please-machine-tickle-Lana'.

3

The Educated Monkey

Sarah learns to 'read' and 'write' in Premack colour-code language • Understands abstract and mathematical concepts • First chimp to master the three Rs.

Soon after Washoe learned how to 'talk' through signs and Lana learned the computer language, Yerkish, another chimp called Sarah was learning to 'read' and 'write' in a third language, called Premack.

Premack is a language system developed by Dr David Premack and his wife, Ann, at the University of California at Santa Barbara. In certain respects it could be said that Lana was either 'talking' or 'writing' with her computer in Yerkish; however, this was a one-way system of communication, with either the machine or the humans responding to Lana's 'written' requests. Premack, like Ameslan, allows a two-way communication.

Premack is a language which employs coloured pieces of plastic with metal backs which are lined up from top to bottom on a magnetic board. When lined up in the correct order, they convey a grammatical construct or sentence that conveys a message.

Through the Premack 'language', Sarah mastered 130 'words' and used them to write sentences and read commands or requests. Among other things, she demonstrated her ability to understand the 'If . . . then . . .' logic conveyed in language (i.e., if you do something, then you will get a reward).

The chimpanzee's ability to 'read' had, to some degree, been proved in earlier experiments; furthermore, it was also accepted that chimps could certainly recognize their own reflections in mirrors and understand abstractions such as photographs and drawings. Lana, with her Yerkish computer, demonstrated her desire to be entertained by 'reading' images on film and not just random images. She had a marked preference for monkey films over human films – she watched a film on chimp anatomy over three hundred times. On one occasion when Sarah was not working on her reading and writing, she was placed in front of a television showing a video about orangutans in the jungle. When one orangutan was captured in a net, Sarah leapt about, hooted and then threw paper scraps – it certainly seemed the chimp equivalent of booing a villainous action.

An exciting aspect of Premack was its ability to teach Sarah abstract ideas such as 'brown colour of chocolate'. These abstractions extended into another area in which Sarah had become proficient: mathematics. Later experiments with Guy Woodruff and David Premack at the University of Pennsylvania demonstrated Sarah's ability to understand mathematical concepts. She developed a sense of mathematical proportions by reading a cup as being a quarter, half, three-quarters full and relating it to a disc of equal proportions. This is basic mathematics and it was taught in exactly the same way that any human child would learn it.

Sarah had become an educated chimp. She had mastered the three Rs: reading, writing and arithmetic.

4

Gorilla Debates

Dr Patterson and Koko the gorilla • Vocabulary of
600 Ameslan 'words' • IQ tests • 'Inventing' words
and 'reading' aloud • Second gorilla, Michael:
telling complex stories.

Since Washoe's initial breakthrough in Ameslan, the most remarkable
developments in 'ape-talk' have occurred in Project Koko. Project
Koko began in 1972 when a Stanford graduate, Francine (Penny)
Patterson, began working with an infant lowland gorilla named
Hanabi-Ko ('Fireworks Child') because she was born on the fourth of
July. Inspired by the work of the Gardners with Washoe, Penny
Patterson's efforts enabled Koko to outdistance all other apes in her
acquisition of language skills. Within ten years the gorilla had
acquired over 600 'words' in Ameslan and habitually used 300 of these
in a working vocabulary.

The key to Dr Patterson's success with Koko is, as she herself
explains, that she has paid considerable attention to aspects which
motivate apes to learn. As Lana pointed out, there is limited
satisfaction in pleasing a computer and the lack of long-term teachers
in nearly all ape experiments does not help their motivation. Koko is
the only language-learning ape who has received continuous
instruction by the same teacher; and the only ape to be kept on such a
programme for more than a few years.

From the ages of four to six, using such devices as the Wechsler
Preschool and Primary Scale of Intelligence, Koko achieved some
remarkable scores in intelligence tests. At four years, on picture
completion tests, Koko scored at the level of a six-year-old human! At
four-and-a-half she outscored six-year-olds in discrimination between
'same and different' in drawings and she also scored remarkably in
problems of logical progression.

Naturally, she did poorly in verbal skills but in picture IQ tests, at the
age of five-and-a-half, she scored as high as 90, which suggested that

she had the same mental capacity as a human child aged four years and eight months.

When it came to word inventions Koko far outpaced the chimps. Dr Patterson recorded dozens of these: 'finger-bracelet' for ring; 'bottle-match' for cigarette-lighter; 'eye-hat' for mask; 'elephant-baby' for Pinocchio doll; 'red-corn-drink' for pomegranate seeds; 'nose-stink' for perfume; 'milk-candy' for tapioca pudding; 'stuck-metal' for magnet; 'lettuce-grass' for parsley.

Among her creative usages are those that have arisen from anger and frustration: 'red-mad-gorilla', 'red-rotten-mad', 'stupid-devil', 'rotten bird', 'stubborn donkey', 'bad-gorilla-nut'. Like the chimps, she also has a measure of scatological usage of made-up curse phrases such as: 'You dirty bad toilet'. When Dr Patterson falsely accused Koko of damaging a toy, the gorilla responded indignantly to her: 'Penny dirty toilet devil'.

Koko spends a fair amount of time 'reading' magazines. Often she will see pictures as she leafs through books and signal 'cat' or 'alligator' as she goes along, much like any child absentmindedly reading aloud. When read aloud to by Penny, she will act out nursery rhymes and read along in sign language.

Another of Koko's activities is playing with dolls. As Dr Patterson herself wrote in *The Education of Koko*:

> Koko spends a good deal of time talking to and playing games with her toys. This is a private pastime of hers and she does not like to be watched while doing it. She seems to get embarrassed.
>
> When she was five, she was playing with her blue and pink gorilla dolls. First she signed 'bad-bad' to the pink doll and 'kiss' to the blue one. Next she signed 'chase-tickle' and hit the two dolls together. Then she joined in a play wrestling match with both dolls. When it ended, she signed 'good gorilla good good'. Finally she looked up and saw she was being watched, and abruptly stopped.

On many other occasions when Koko has been observed 'talking' to her dolls she has also been seen to shape their hands in such a way as to actually 'teach' them how to make signs for such words as 'eat' and 'more'.

After a time, Dr Patterson acquired a second gorilla, a male called Michael who was two years younger than Koko. After a few years of training, Michael also became proficient in Ameslan. This led to gorilla conversations and sometimes to arguments. On one heated exchange, Koko called Michael a 'toilet-devil'. The younger Michael didn't take this passively; he responded with the considerable invective: 'Stink-bad squash gorilla lip'.

Arguments about animals only being able to perceive immediate experiences, rather than to remember past events, were dismissed after a number of incidents occurred in which the gorillas were observed telling tales about happenings in the past. On one very memorable occasion, a trainer assumed that Michael was just having a bizarre flight of imagination when he recounted an extraordinary story about a red-haired girl. Using the phrases 'hair red girl', 'hit in mouth red bite' and 'girl big trouble', the gorilla was obviously attempting to convey a message of some urgency to the trainer. The trainer duly recorded Michael's statements but could make no sense of them. Later it emerged that Michael had been considerably traumatised by an incident in which a red-haired girl had stormed into the trailer of a neighbouring lab assistant. A violent argument and fight had broken out, which ended only when the police physically subdued the woman at gunpoint.

Michael also seemed somewhat disturbed by the laboratory cat's primitive hunting habits. On at least two occasions he had found it necessary to tell tales on him. 'Cat cat chase eat bird . . . cat hit bad . . . cat bird frown hit-in-mouth'. And some time later, almost like a little short story: 'bird good cat chase eat red trouble cat eat bird'.

Strangely enough, for supposedly fierce jungle beasts, both Michael and Koko seemed to be disturbed by any sort of violent action. And both seemed to possess considerable narrative skill when it came to telling stories, which often had the moral overtones of committed pacifists.

5

Planet of the Apes

Carl Sagan's speculations • Washoe's ability to
teach sign language •. Comparing brains of
humans and apes • Darwin's theory of language
and brain growth • A nation of apes?

In his book, *The Dragons of Eden*, Carl Sagan speculated on the
possible outcome of a large breeding colony of chimpanzees that were
taught sign language and allowed to evolve for a few centuries. He
concluded that 'the long-term significance of teaching language to the
other primates is difficult to overestimate'.

It is now well-known that apes in the wild 'teach' their children and
that different tribes of apes have different 'cultural' habits. In captivity,
the chimp Washoe has 'taught' some sign language to her adopted
infant Loulis; while observation of Koko's behaviour with her dolls
and her 'conversations' and 'instruction' to Michael lead us to the
conclusion that she would certainly teach Ameslan to any offspring
she might have.

It has been observed that basic English can be achieved with as few
as a thousand words and in less than one ape's lifetime a working
vocabulary of over a third of this has been achieved. What might be
possible in several generations' time, and what evolutionary factors
might such a gift of language trigger off in the apes?

Oxford psychologist Dr Richard Passingham once wrote in an article
on the chimpanzee:

> Biologists have established the extraordinary genetic similarity of
> man and chimpanzee. The resemblance between non-repeated
> sequences of DNA in man and chimpanzee is greater than that
> between mouse and rat; and the average protein differs less than one
> per cent between the two species. If the genetic distance is so small,
> why is the mental gap so great?
>
> The secret must lie in the brain. The brain is the only one of the
> human internal organs that is larger than expected for a primate of
> our size . . .

At birth the brain of a baby is little bigger than we would expect for a newborn chimpanzee if it were the same body weight. Before birth the brain grows at the same rate in man as in chimpanzee and macaque monkey. But in the macaque the rate slows down markedly just before birth, whereas in man the brain continues to grow at the very rapid rate that is characteristic of the foetus for two years after birth. It needs only a small modification in the control genes that direct the pace of growth to come up with a brain as large as ours.

This is a startling thought. It wouldn't be a human brain, of course, but it would make for a highly intelligent chimpanzee. And from what we already know about genetic engineering, it seems likely that such an experiment will be possible in the very near future.

But genetic tinkering and the startling aspect of a chimpanzee-Frankenstein apart, this is perhaps looking at the evidence backwards. From what we know of evolution, man did not suddenly appear with a huge brain which made complex thought and language possible. Man's ancestors began with a brain at least as humble as the chimpanzee's but which evolved into something far greater. So what factors caused the evolution of the human brain (or, in Dr Passingham's terms, what element in the life of man's ancestors led to that 'small modification in the control genes' that permitted the growth of the brain)? Could these same factors result in the rapid evolution of the ape brain? Charles Darwin, the father of evolution, seems to have no doubt on the matter. Darwin perceived that it was the factor of language that facilitated the growth in both the size of the brain and the intellect in humans, and that such a factor might have a similar effect in apes.

> The difference in mind between man and the higher animals, certainly is one of degree and not kind. If it could be proved that certain high mental powers, such as the formation of general concepts, self-consciousness, et cetera, were absolutely peculiar to man, which seems extremely doubtful, it is not improbable that these qualities are merely the incidental results of other highly advanced intellectual faculties; and these again mainly the results of the continued use of a perfect language.

In short, Darwin concluded that humans were not biologically unique animals with freakishly large brains but that they had evolved their large brains through natural selection and that language had played a major part in the evolutionary process.

Having now introduced the language factor to apes, it seems quite possible that, permitted a small 'country' of their own which was safe from our human excesses, colonies of language-using apes might very well come up with some interesting answers to Sagan's rhetorical question: What sort of culture, what kind of oral tradition would chimpanzees establish after a few hundred or a few thousand years of communal gestural language?

Is it possible that, in time, another primate species might evolve from this nation of apes – a species that might even eventually rival our own domination of the planet?

6

Whale Songs

Songs of the humpback whale • Roger and Katherine Payne, whale musicologists • Mystery of their songs • Structures, dialects • Changing themes and compositions.

In 1970, the largest single pressing of any phonographic record in history was made. Capitol Records, in co-operation with National Geographic, produced ten million copies. Remarkably, the recording artists who merited this mass publication were totally unknown to the public. Even more remarkably, the singers were not human. Most astounding of all, the record was a smash hit. The humpback whale had overnight become the world's greatest recording artist.

The recordings of the 'Songs of the Humpback Whale' were made by the world's foremost whale musicologist, Dr Roger Payne. Dr Payne was a neurophysiologist. He first studied owls at Cornell University and then moved on to study whales. Payne also had an extensive musical background, and he became fascinated with the study of the 'songs' of the whales.

The humpback whale is the greatest 'singer' among the great whales. It is called 'humpback' not because of some implied deformity but because of its tendency to arch its back when it dives deep. It is capable of spectacular leaping and when viewed underwater proves to be sinewy and agile using its fourteen foot pectoral flippers to steer and soar its way through the deep waters 'like swallows on wing', as one old sea captain remarked.

Why humpback whales should sing is not really known but, as scientists and musicians, Roger Payne and his wife Katherine found themselves drawn further and further into the study of their eerie symphonies. Obviously, 'whale song' is some form of communication but it seems to be many other things as well. As Katherine Payne said at a symposium in San Francisco:

> The humpback is the only known animal species whose songs are continuously undergoing changes, changes that are complex, large scale and rapid.
>
> The whales use a technique very much like one a good composer uses to create beautiful and interesting music. For example, Beethoven sets up rhythmic patterns and musical themes we all recognize and expect to recur. Then he surprises you with a variation. Every humpback whale song I've heard surprises me just the way Beethoven does.

The whales sing songs in solo, in duets, trios, quartets and even, in entire schools, in chorus. Whale songs are like human dialects. All whales in one place will sing one dialect, while their cousins in another breeding ground will sing a completely different arrangement.

Songs change and evolve very gradually over weeks and months, in length, in timing and in number of sounds. Dr Payne equates these changes with musical 'themes' in symphonies, or to use a simpler analogy, with songs with 'musical rounds' which are dropped out, split apart and rearranged. After a time, all phrasing is gradually replaced and a completely new song emerges.

'The songs and the changes give us a real handle on how whales think and a way to study the process of cultural transmission in a species other than man.'

7

Voices from Atlantis

**Echo-location in whales and dolphins • Dr Lilly's
experiments with dolphins • Computer analysis of
'dolphinese' • Cetacean brains and intelligence •
Difficulties of translation.**

Investigation of whale and dolphin communication seems to have
largely arisen after allied navy engineers began using underwater
listening devices to track submarines. The engineers' headphones were
often flooded with whole concerts of bizarre sounds: chattering,
clicking, burbling, clunking, rattling and creaking. These were the
sounds of whales and dolphins. Later came the complex but definitely
musical sounds of the humpback whale and the impossibly powerful,
low-frequency, transocean sounds of the fin and blue whales which, to
the human ear, are more like 'sensing an earthquake' than hearing.

In Florida, in 1946, a researcher named Arthur McBride discovered
that one purpose of the sounds created by dolphins and all toothed
whales is that of 'echo-location' or the use of underwater sound as a
radar system. This enables dolphins, for instance, to catch fish and
avoid nets even in the dark. Later research revealed that the clicking
sounds of dolphins originate in a soft tissue area in the forehead,
called the melon, and the signal is received in the lower jaw which acts
like an internal antenna. Among the many mysteries of this process,
however, is the question of where these sounds came from – or rather,
how they are produced. Whales and dolphins have no vocal chords and
there is no visible muscle or bone that is mechanically capable of
making such sounds.

All this was of interest to only a handful of scientists until Dr John
Lilly published his book *Communications Between Man and Dolphin*
in 1961. It was Dr Lilly who drew the public's attention to the concept
of 'interspecies communication' with his studies of dolphins and
whales.

In the mid-fifties it was discovered that a dolphin could
'communicate' with humans like many other trained animals. It could

come to the surface and whistle and make strange sounds through its blow-hole. However, by 1957 dolphins were also seen to be demonstrating an ability to mimic human speech.

Lilly stunned many of his colleagues by saying: 'These cetaceans with huge brains are *more* intelligent than any man or woman.' Many scientists reacted as if it were their duty to defend the pride of the human race and immediately attempted to discredit Lilly's research. Others who chose to work with larger cetaceans, however, have come to conclusions which tend to support his ideas to a certain degree. In 1965 Thomas Pulter, of Stanford Research Institute and director of its Biological Sonar Laboratory, began for the first time to observe whales in a systematic way when a killer whale named Namu was kept in the public aquarium in Seattle. He discovered 'conversations' between Namu and whales seven miles away from the aquarium. He described a very complex exchange of sounds that could only be described as an 'animal language'.

By 1967, Dr Lilly refused to conduct any further captive dolphin experiments. This, he insisted, was a moral decision. That year he released his dolphins on the grounds that 'I no longer wanted to run a concentration camp for my friends'. What he did do, however, was continue to work with complex computer analyses of dolphin 'language', in an attempt at code-breaking what he calls 'dolphinese', his ultimate goal being to invent a machine that simultaneously translates conversations between men and dolphins. Lilly called this his 'Janus system' experiment – named after the Roman god with two heads who is guardian of the gateway; in this case, one head is a dolphin and the other is human.

Lilly's experimental work has shown that there is a fundamental difference in 'interspecies communication' when cetaceans are compared to apes. Whereas it is necessary to 'educate' apes by teaching them a form of language, with whales and dolphins the evidence suggests that a language already exists and it is a matter of finding a way of translating it. In fact, Lilly insists that having been the ruling intelligence on this planet much longer than humans, whales and dolphins have evolved a superior 'civilization'. They are symbolically the inhabitants of Atlantis, that mythical civilization beneath the sea.

Carl Sagan once speculated: 'The brain of a mature sperm whale is almost 9000 grammes, six and a half times that of the average man . . . What does the whale do with so massive a brain? Are there thoughts, insights, arts, sciences and legends of the sperm whale?'

Dr Lilly's writings have brought about a widespread belief in the possibility of establishing a communication link with cetaceans. However, anyone involved in such experiments understands the difficulty of the translation process. As Dr Kenneth Norris of Santa Cruz, California, has observed, dolphins and whales 'see' and 'taste' through sound and possess many other faculties of which we are now only vaguely aware: 'It is very hard for us to imagine sensory systems and processes we do not have . . . It's a bit like a man from outer space tapping into the Bell System center and trying to make sense of all the beeps and switching sounds.'

But slow-going though it may prove, study of the large and complex cetacean brain convinces most scientists of the animals' intelligence and capacity for dialogue. It is in the seas of our own world that many believe humanity has its best chance of communicating with an alien intelligence, not out in the stars hundreds of light years away.

8

Is Anybody Listening?

Hostility towards interspecies communication experiments • Implications are a moral minefield for scientific community • Communicators are consorting with the enemy • Non-human intelligence and the spectre of animal liberation.

Since the remarkable breakthroughs in the field of interspecies communication during the sixties, there has been widespread interest from the general public in experiments of scientists like Dr Patterson and Dr Lilly, and excitement at their potential. However, as Ted Crail points out in his book *Apetalk and Whalespeak*, there is considerable hostility within the scientific community itself towards the advances that have been made.

It is more than mistrust of the methods and goals of this research that disturbs scientists and provokes them to try and discredit the work of their colleagues. One reason for this seems to be a kind of righteous indignation somewhat reminiscent of that often-quoted bishop's wife's response to the theories of Darwin: 'Descended from monkeys? My dear, let us hope that isn't true! But if it is true, let us hope that it doesn't become widely known!'

Among scientists, there seems to be a need to defend man as the only language-using animal. At nearly every stage of ape or dolphin communication, there have been objections that these animals are not capable of acquiring 'language' as we know it. First, it was argued, they were incapable of forming sentences, or abstract concepts. Then it was: logical progression, or mathematical constructs, or initiating dialogue. At each stage the animals proved the critics wrong. As for the translation of a cetacean language, this was said to smack of mysticism and (horror!) parapsychology.

Beyond basic human indignation at having man's status undermined as the only species capable of using language, there is a more specific threat to the scientific community implicit in research in interspecies communication.

The truth of the matter is that a large faction of the scientific world has a vested interest in emphasizing the difference between humans and animals, especially with regard to intellectual capacity. If we attribute intelligence to cetaceans and apes, even if we only put them on the same level as a human five-year-old which now seems a modest claim – where does that leave the scientists who conduct painful and fatal experiments on these animals?

It makes the moral propositions of the animal liberation groups almost impossible to ignore by that sector of the scientific community that makes its living from animal experimentation. Furthermore, it might prove very difficult to conduct an experiment on a chimpanzee who is attempting to talk you out of it. A case in point is the unhappy Ameslan-taught monkey Nim Chimpski, who, when shifted from one owner and cage to another, kept signing 'Want out'.

Scientists who were once demonstrating the fact that the animal knew what it was saying when under language instruction, were now claiming that animal responses were just habitual signs and of no significance. The confused chimp must have felt that humans had suddenly gone deaf or – in the case of Ameslan – selectively blind.

Breaking down the barriers between 'them' and 'us' is one of the first steps towards peace in any kind of war. Yet there is a certain element in the scientific world that sees the establishment of a dialogue with the other side as tantamount to treason. If pain, suffering, happiness, love, are found in equal measure in 'them', what is it that makes us uniquely 'us'? Even more, what gives 'us' the moral imperative to deny 'them' the basic rights of all thinking, feeling, human animals? The whole issue of animal rights is turned on its head.

One way round the problem is to deny and sabotage any attempt to establish the communicative link in the first place, and to discredit any evidence that suggests these mute beings think and feel as we do.

We have been here before, of course, in the debate over human slavery, as any animal rights advocate will tell you. In certain parts of the world you can still hear this debate in progress today. Racism and specism share a common desire to maximize the uniqueness and apartness of 'us' and 'them'.

Jeremy Bentham, as a radical campaigner against slavery, brought up this very issue in the eighteenth century:

> The day may come when the rest of animal creation may acquire those rights which never could have been withheld from them but by the hand of tyranny . . . the question is not can they reason? nor can they talk? but, can they suffer?

In fact, the Communication Corps in this ecology war has proved that non-human animals are capable of all three.

Because of the potential problems raised by 'talking animals', many experiments in this direction have been curtailed. In the cases of Dr Patterson and Dr Lilly, rejected by the scientific community, they have had to seek public funding to continue their research.

Other scientists working in the field of animal communication, however, have received large sums of government money because they have seen a marketable potential in 'educating' animals. That potential is inevitably military.

Rather than spending money on animal experiments that give us a glimpse of the distant peaceful Eden, fortunes are being spent to bring animals into the hell of our wars. In this interest there has been no end to efforts and funds.

6

Animal
Soldiers

The uses of animals
in warfare

'No sooner does man discover intelligence than he tries to involve it in his
own stupidity.' *Jacques Cousteau*

1

Doves of War

Monument to the unknown pigeon • 20,000 dead
pigeons • 'Cher Ami', the hero pigeon • Microfilm
and pigeon aerial photography.

In a municipal park in the French textile manufacturing town of Lille,
there stands one of the world's most unusual monuments to the war
dead. It is not very different architecturally from the innumerable
other monuments to the European victims of the two world wars.
What makes it unique is the nature of the twenty thousand war dead it
commemorates as fallen heroes. The names of the fallen have never
appeared on any official casualty lists of any Allied or Axis powers. The
reason for their anonymity is that these soldiers were not men but
pigeons.

In the course of the First and Second World Wars, some twenty
thousand carrier pigeons were killed serving as couriers for the Allied
armies of Europe. This monument in Lille was built to commemorate
these sacrificed doves of war.

Many are the stories of men and armies who have owed their
survival to the success or failure of carrier pigeons making their way
through enemy lines. Among the most famous in the First World War
was the pigeon called 'Cher Ami '. This bird sustained seven wounds
yet still managed to deliver a vital mesage before expiring. In doing so,
she saved New York's 77th Division – later called the 'Lost Battalion '
– from certain annihilation in its eleventh hour.

Upon her death, 'Cher Ami' was lovingly embalmed and is now
enshrined in the American Smithsonian Institute in Washington, a
small feathered monument to an episode in American history.

She is one of the two most famous pigeons in American history. The
other pigeon, called 'Martha', who ironically died in the first week of
the same war, has also found a last resting-place in the Smithsonian.
However, she was the victim of a different kind of war – a war of
extermination. In the course of fifty years, American market hunters
killed thousands of millions of passenger pigeons, the most numerous

bird species on the planet. 'Martha' was the last passenger pigeon the world has ever seen.

Pigeons have had a long history as military couriers, dating back to ancient times. However, by the time of the Franco-Prussian war technology was such that, with the use of microphotography, a single pigeon was capable of carrying 40,000 messages. During the Prussian siege of Paris in 1870, pigeons carried literally millions of letters and dispatches out of the capital.

Photography offered another role for the pigeon. The first aerial photographs ever taken were by pigeons and before the outbreak of the First World War the German photographer, Julius Neubronner, had aerial pigeon-photography down to a fine art. He had developed a miniature camera weighing $2\frac{1}{2}$ ounces, which produced $1\frac{1}{2}$ inch square negatives that made exposures timed at thirty-second intervals. Pigeons had become the first eye-in-the-sky espionage agents.

2

B. F. Skinner and the Pigeon-Guided Missile

Project Orcon: Skinner's pigeons in the nose-cones of Pelican missiles as guidance systems • Vietnam use of 'stool-pigeons' with microtransmitters • Vulture post and other systems.

By the time of the Second World War, new and more aggressive roles for pigeon soldiers were found. Possibly the most bizarre scenario in the drama of warrior doves was that involving the father of behaviourist psychology and the inventor of the rat maze, B. F. Skinner.

The war economy seemed to have a place for everybody, and so B. F. Skinner and the pigeon were fatefully brought together to provide a solution for the problematical US Navy's 'Pelican'. The Pelican was not a water bird, but in fact America's rather ungainly answer to the infamous German V-rockets. The problem with the Pelican was that it had no sense of balance and very little sense of direction – two very considerable drawbacks for a supposedly 'guided' missile.

B. F. Skinner believed he had found a way of stabilizing the clumsy Pelican. So, in June 1943, he was contracted to develop a guidance system for the missile. This research was later continued at the Naval Research Laboratory in Washington under the name Project Orcon ('Organic Control'). Skinner's solution was simplicity itself: to put a well-balanced homing pigeon in the nose-cone of the missile. The problem, Skinner concluded, would be resolved by the fact that, properly wired, the balanced inner ear mechanism of the pigeon would always cause the bird to remain upright. Meanwhile the direction of the missile could be corrected by continuous pecking by the bird at a target image projected through a lens system.

It was not such a great solution for the pigeons involved, of course, as they sat with several tons of explosives packed in behind them. However, Skinner reasoned, everybody had to pull his weight in the war effort, even pigeons.

Experiments simulating the 660-mph missile dive on a target ship demonstrated fairly conclusively that the pigeon could act with considerable accuracy as a guidance system. However, despite the success of the test trials, the system was not put to practical use against the German or Japanese navies. This, Skinner sadly concluded, was because of military officers' prejudice against trusting the war to a bird brain. Somehow, it was felt to be un-American and rather low-tech.

Among those working with B. F. Skinner on the pigeon-guided missile system at the University of Minnesota in the early 1940s were Keller Breland and his wife Marian, a husband-and-wife psychology team. And although Skinner left the pigeon behind him with the war, the Brelands went on to found Animal Behavioral Enterprises in 1947. 'They took my theory, applied it well enough to make a living from it and, in the process, had all sorts of useful accidents which we've all learned from,' said Skinner.

The Brelands and others, like Robert E. Lubow – a psychologist working in the same field in the 1960s – found the government a constant source of funds when it came to 'applied pigeon psychology for the military'. During the Vietnam War, for instance, the American Air Force commissioned research into 'Airborne Biological

Reconnaissance Systems'. In layman's terms, this was the pigeon spy or 'stool-pigeon': doves who were trained to detect ambushes in the jungle. The idea was that the stool-pigeon (fitted with a microtransmitter) flew ahead of patrols and, upon spotting any enemy soldiers hiding in the undergrowth, could be relied upon to make for the nearest tree and remain there cooing and pointing out the guerillas.

In preliminary experiments and field work, the birds seemed to perform well; however, time alone would tell how well they operated in jungle conditions. Sadly the 'Airborne Biological Reconnaissance Systems' were entirely wiped out by the combination of a form of South-East Asian pigeon-pox and the fact that the Vietnamese villagers found the birds very good eating and shot all pigeons on sight.

Experimenters did not work solely with 'airborne pigeon systems'. Inevitably, the need arose for a method of conveying rather more bulky packets to soldiers trapped in jungle war zones, and numerous experiments were conducted by Dr Lobow using starlings, crows, ducks, geese, and finally, turkey vultures, to demonstrate a capacity for delivering much heavier loads. Once again, the reason for the lack of widespread use of these new 'airborne systems' seems to have had more to do with human psychology than its practicality. Consider how the morale of a surrounded and entrenched troop of soldiers in a combat zone might have been affected if central command's messages were constantly conveyed by a vulture! Useful as it may have proved, the vulture-post system was a dead letter.

3

Dr Napalm and the Incendiary Bat

Dr Feisner's bat bombs • US experiments •
Premature cremations • Loss of $2 million hangar •
Frozen bat bombs.

Birds were not the only flying creatures drafted into military service. In 1943, Dr Louis Feiser, a Harvard professor and the inventor of napalm, was brought in to perfect a new weapon in the US Army arsenal: the Incendiary Bat. To this end the US Army Corps captured thousands of bats from the Carlsbad Caverns, New Mexico, and allocated some two million dollars towards the development of the bat-bomb. Dr Feiser developed an incendiary device that weighed about one ounce, which was surgically attached to the bat.

The plan was that thousands of bats would be airdropped over Japanese cities with these miniature fire bombs sutured on to them. The bats would immediately seek refuge in attics and within various public and private buildings. There they would chew through the suturing string and trigger off the delayed-action fuse which would cause the fire-bomb to ignite and destroy the building.

However, despite Dr Feiser's successful development of the explosive devices, the bats proved somewhat troublesome creatures. A frequent problem was that they chewed through the surgically attached strings too soon, thus igniting the fire-bomb before it had dropped. In one case, this resulted in the burning down of a two-million dollar aircraft hangar where experiments with the bats were taking place, and in another, the fire-bombing of a general's car.

Changing tactics, it occurred to the scientists that when bats are exposed to cool temperatures they go into a sleeping coma akin to hibernation. In this condition, they would be less inclined to chew through the bomb fuse-cord. To this end, an experiment was launched with the bat bombs. Crates of cooled-out bats were taken up in an aircraft, then, at a sufficiently high altitude above the New Mexican desert, they were methodically dropped while in their sleeping state.

The idea was that the slumbering bats would awaken in their long fall through the desert air of New Mexico. Instead, the flat desert land of New Mexico experienced a most peculiar weather condition : hundreds of frozen, sleeping bats rained down like huge, oddly-shaped hailstones, thudding and bouncing off the hard earth.

Despite this fiasco, the experiments continued for another two years. The project was dropped at one time by the army, then picked up again by the navy. However, by the time the incendiary bat-bomb was finally declared operational, the atomic bomb was on the way, making these miniature bombers redundant.

4

Bazooka Dogs and Torpedo Hounds

Anti-tank dogs with explosive packs used by Russians • Vietnam US dog soldiers • Navy experiments with dogs in submarine torpedo tubes.

After horses, dogs are probably the most common auxiliary force attached to human armies. From the time of the first tribal peoples, men have taken their pet hounds into the fray of battle with them. In ancient and medieval times, some war hounds were even equipped with spiked collars and armour; they often harrassed or even killed enemy troops and horses.

In the First World War, dogs were primarily used as sentries, carriers of munitions, and tracking animals. The French Army had a particularly reliable network of dog messengers between the front line troops and officers in command and consequently suffered more than five thousand casualties among its dog soldiers.

It was the Russians who found the most deadly use for dogs in warfare. Using Pavlovian methods, they trained dogs loaded with extremely highly-charged explosives to run under the belly of enemy tanks. The explosive pack on the dog's back had an extended antenna which, when it struck the metal underside of the tank, triggered the detonator destroying the tank – and not doing the dog very much good, either.

During the Vietnam War, 'stalking dogs' were used to hunt out enemy troops. Others were trained at the Land Warfare Laboratory near Washington DC, to be used as long-range sensors or scouts, sniffing out mines, trip-wires and dead or wounded soldiers. So useful have dogs proved to be that the American 82nd Airborne Division at Fort Bragg, North Carolina, has even experimented with training them as parachutists, in order to provide back-up for their advance troops.

Rather late in the Vietnam conflict, something of a scandal developed when it was learned that, in all cases, Vietnam dog soldiers were getting one-way tickets to the war. If they survived the battlefield it emerged that they were 'retired'. That is, they were either destroyed or turned over to the Vietnamese, who used them as a source of protein rather than companionship.

Mindful of human sympathies towards dogs, most canine war experiments were kept secret for many years after they were attempted. One such secret experiment in submarine warfare took place in 1942. Just a year before B. F. Skinner began his pigeon-guided missile system, a similar system was being considered exploiting the dog's supersensitive hearing system to pick up the pinging sound of submarine sonar. The plan was to place a dog in the hollow nose of the torpedo and fire it off at the enemy. Fortunately for the dogs, after a certain amount of testing, this scheme was shelved in favour of newly developed mechanical systems.

5

The Minesweeper Pig

Dr Lubow's explosives-detecting pigs • Morale
problems • Pig offensive: Dr Bailey's pig bombs.

After his experiments with various 'airborne biological systems', Dr
Robert Lubow turned his attention to more earthbound 'biological
systems', particularly to 'explosive detector systems'. This military
jargon usually meant dogs who were trained to sniff out explosives.

Dr Lubow, as demonstrated by his experiments with turkey vultures,
refused to be bound by tradition. Consequently, he experimented with
a number of animals, all of which he believed achieved better results
than dogs. Some of the animals tested were racoons, coyotes, cats,
foxes, javelins, civets, ferrets and even skunks. However, the hands-
down winner over all of these was the domestic red duroc pig.

It was found that even the best-trained mine-dog could only smell
out explosives buried a few inches beneath the surface, whereas pigs
were capable of sniffing out bombs buried over a foot below ground.
However, despite the indisputable proof of the effectiveness of his
minesweeper pigs, that same prejudice that destroyed the chances of
the turkey vulture courier and the pigeon-guided missile seems to have
blighted Dr Lubow's newest discovery. No reason has been given for
not using the minesweeper pig in place of dogs, either in military or in
civil situations. However, it is believed that some soldiers feel there is
something a little undignified in being a pig trainer. In civil situations,
it seems that most police would rather take their chances with a dog
(even if they suffer the occasional casualty), than be seen walking into
a suspected bomb area with a pig on a leash.

Undeterred by the unacceptability of the pig as a part of man's
defensive arsenal, a number of experiments have recently developed
the pig as a potential offensive weapon.

After the Russian bazooka dog and the American napalm bat, it was
not altogether surprising when the *New York Times* revealed that the
CIA was training dogs, cats, seals and otters to carry explosives and

microphones for either sabotage or espionage purposes. However, the pig bomb concept is something rather different, and a number of researchers have boasted innovations in this area. Dr Marion Breland-Bailey of Animal Behavior Enterprises in Hot Springs, Arkansas, under military contract, actually cut open a number of wild boar and implanted large metal and plastic objects, coated with beeswax so that the animals' bodies could tolerate them more easily. The experiments demonstrated that pigs can carry up to twenty-three pounds of explosives internally.

The advantage of this system was that unless an enemy actually checked each pig for unsightly surgical scars, there would be little likelihood of the pig-bomb being detected in a rural setting. The swine could be trained to infiltrate enemy barracks or encampments and could be detonated at will.

6

Elephant Paratroopers and Kamikaze Cats

US parachute elephants into Vietnam jungle combat zones • Experiments with cats being dropped with explosive charges from aircraft • Recent bio-cybernetic experiments.

The trouble with weapons like the pig-bomb is that it often proves to be very bad news for the rest of the species. After the initial shock of surprise, the response of the enemy will be: shoot all pigs.

The American themselves amply demonstrated this in Vietnam when, even without the stigma of a capacity for exploding, all elephants and water buffalo that were 'suspected' of helping to carry supplies for the Viet Cong were machine-gunned by helicopter gunships. Even cattle and poultry 'suspected' of being eaten by the Viet Cong were similarly slaughtered. Entire forests came under 'suspicion' as wilful collaborators, and were consequently fire-bombed and pulverized into swamplands. As one American general pointed out with almost unbelievable understatement in the midst of the Vietnam campaign: 'Soldiers can't be expected to be conservationists.'

The American military in Vietnam, however, sometimes had cause to regret its indiscriminate killing of animals. After going into military areas and shooting, for instance, the 'suspect' elephant collaborators, it occasionally found itself back in those same areas and in need of heavy transport animals for its South Vietnamese allies. Nothing daunting American ingenuity, the Defense Department developed an elephant paratrooper division whereby it parachuted tranquilized elephants into the combat zone for its own use.

When news of this latest initiative eventually reached the public in late 1968, the response was alternately bemused and angry. In Britain, in particular, the action was seen as absurd and inhumane. However, one American military attaché at the US Embassy in London, when confronted by concerned animal supporters, quite blithely and with evident sincerity answered that the cruelty charge was untrue. Quite apart from being tranquillized, he said, sending the animals by air and dropping them by parachute was in fact a great kindness. It obviously spared the animals a long walk through the jungle.

The furore over the paratrooper elephants only served to prove to the American military that the less the public knows about its activities, the better. Little concern was shown during the Second World War, for instance, with experimentation on the 'Kamikaze Cat' – and with good reason. No one knew about it until decades later.

The Kamikaze Cat concept was the brainchild of the wartime Office of Strategic Services (OSS) – the forerunner of the CIA – which was set up under President Roosevelt with the guidance of British Intelligence. The experiment was based on the commonly observed aversion that cats have to water. It seems to have evolved through the following logical progression:

Problem: Japanese Kamikaze flying bombs have a certain advantage over conventional American bomber squads. Given that the current political climate in America does not facilitate a similar squad of men in our service, another method might be devised.

Observation: Cats will go to any lengths to avoid water.

*Proposal:*Would it not seem logical that when jettisoning high explosives from aircraft on targets such as ships at sea, if a cat were attached as an extra guidance system, the cat's terror of water and sure balance would in all likelihood result in the cat-bomb landing on the deck of the target ship?

As most thinking individuals might conclude, this experiment was destined to go the way of the vulture courier. After numerous experiments in which defenceless cats were slung out of aircraft, it was discovered that fear of wet paws is not uppermost in the mind of a cat when hurtling through space with a bomb strapped to its back.

Incredible though it may seem, the concept of the Kamikaze Cat has not entirely lost its appeal in some circles. The Soviets are apparently working on an updated version of it for their missiles. The RAND Corporation's *Soviet Cybernetic Review* contends that the Russians are 'experimenting with disembodied cats' brains . . . creating bio-cybernetic packages for implantation in air-to-air missiles. The cybernized cats' brains will, if all goes as planned, be able to recognize optical impulses emanating from their targets and to transmit guidance signals accordingly, so that the missiles always stay on target.'

7

Crocodile Battalions

Idi Amin's 'crocodile capital of the world' • Executioner crocodiles of Uganda • Estuarian crocodile • Ramree Island massacre: death of one thousand Japanese soldiers.

In most respects, Idi Amin was as much of a disaster for the wildlife of Uganda as he was for the human population. His troops, with the aid of helicopter gunships, massacred elephant herds for their ivory, shot fur-bearing animals for their skins and killed hoofed animals for meat or just for target practice. Yet, perversely, it could be said that Idi Amin was the champion and saviour of one species. Under his expert hand the Nile crocodile actually hugely increased in numbers.

During the nineteenth century, there was a famous Nile crocodile called Lutembe that lived in a small bay in the Murchison Gulf in Lake Victoria. Coming ashore when bidden by the kings of Uganda, this fourteen-foot beast served for over fifty years as the royal executioner.

It appears Idi Amin revived this ancient tradition on a massive scale. Crocodiles provided Amin with a convenient means of disposing of tens, and perhaps even hundreds, of thousands of bodies. In an astonishing French documentary film, *Idi Amin Dada*, the Ugandan president took a film crew on a journey up the Nile in his private river boat. Showing the interviewers the thousands of well-fed crocodiles lying sunning themselves on the banks of the river, he called out to them. The animals immediately slid into the water and swam expectantly round the boat.

'See,' he said proudly, 'they know me.' Then with a broad gesture and a wide smile, 'This is the crocodile capital of the world.'

Although proper statistics are not kept, crocodiles – of all large predators – are by far the greatest killers of human beings. Idi Amin's efforts aside, crocodiles are probably responsible for at least three thousand deaths annually. The two most dangerous species of man-eating crocodiles are Amin's Nile crocodiles and the even more lethal saltwater Estuarian crocodiles – an animal that has been measured at a record length of twenty-eight feet and has an estimated weight of three tons.

The Estuarian crocodile is the world's largest living reptile, and on one occasion this crocdile has joined in a military conflict which must go down in history as the single greatest mass killing of humans by vertebrate animals.

This strange, and admittedly accidental, partnership of the British Army and the Estuarian crocodile took place on 19 February 1945. On that fateful day, the British had engaged in a battle with an army of Japanese infantrymen on Ramree Island off the west coast of Burma. The Japanese were trapped between a division of the British Army on one side and the Royal Navy on the other. During the day they suffered heavy attack from the army and terrible shelling from the Royal Fleet guns. Over a thousand Japanese infantrymen fled into their only avenue of escape, the eighteen miles of mangrove swamp between the southern inshore edge of the island and the Burmese mainland. This area was entirely waist-deep water and mangrove jungle without any dry land.

The swamp proved to be the prime habitat of the Estuarian crocodile. Although the shelling during the day and early evening had caused the majority of crocodiles to flee the area, the stillness of the evening and the scent of blood in the water brought them back in huge numbers. Roger Caras, in his book *Dangerous to Man*, quoted naturalist Bruce Wright's eyewitness account:

> The din of the barrage had caused all crocodiles within miles to slide into the water and lie with only their eyes above, watchfully alert. When it subsided the ebbing tide brought to them more strongly and in greater volume than they had ever known it before the scent and taste that aroused them as nothing else could – the smell of blood. Silently each snout turned into the current, and the great tails began to weave from side to side.

The terrible screaming of the Japanese soldiers and the deep roaring of the feeding crocodiles was clearly audible to the British troops anchored outside the swamp all through that long night.

Of the thousand Japanese who fled into the swamp, just twenty survived to surrender thankfully the following day.

8

Shark Fleets and Dolphin Squads

Shark attacks • WW II: US cruiser, *Indianapolis* • British steamer, Nova Scotia, torpedoed – 700 lost to sharks • Dolphin couriers and spies • US and Russian use of dolphins • Anti-dolphin dolphins • Killer-dolphins: 60 human deaths.

Although the deadly combination of the British Army and the Estuarian crocodile resulted in the greatest mass killing of men by animals on record, it was a freakish incident, unlikely to be repeated. A much more common murderous combination was that of the submarine and the man-eating shark. Under normal circumstances sharks, despite their reputation, kill about a third as many people as crocodiles – possibly a thousand a year. However, since the advent of submarine warfare there have been hundreds of incidents (especially in the Japanese-American war in the Pacific) when sharks came in for the kill.

One of these occurred only a few months after the crocodile attack on Ramree Island when the US cruiser *Indianapolis* was torpedoed by a Japanese submarine in the Pacific. By the time the rescuers arrived, two-thirds of the crew had either drowned or been torn to pieces by sharks. It was a killing frenzy that turned the waters red with human blood.

But the greatest single mass killing by sharks took place a few years earlier, on 28 November 1942, and it was almost a match for the Ramree massacre. This happened as a result of the torpedoing of the Liverpool steamer *Nova Scotia* by a German U-boat some thirty miles off Zululand in South Africa. There were 900 passengers on board when the ship was hit, but when a Portuguese rescue ship arrived many hours later it gathered only 192 survivors. Over 700 people had either been drowned or devoured by the frenzied, feeding sharks.

In situations of human warfare sharks have proved very efficient but not very discriminating. With little in the way of a brain, these murderous machines would as soon eat a Japanese as an American, a communist as a capitalist, thus presenting certain strategic problems. There are other animals of the sea that would be far more dependable if they could be properly trained. The dolphin, of course, is the most obvious example.

Early financing of research into communication with dolphins was motivated in military circles by the belief that these animals could be exploited in marine warfare and, since 1965, both the Americans and the Russians have been training dolphins with this end in mind. In the US that year a dolphin named Tuffy was trained to carry messages and tools two hundred feet down to the aquanauts in the *Sealab II* capsules. Elsewhere dolphins were soon being used to carry espionage and surveillance devices, and to recover equipment and explosives from the sea-bed. From the mid-1960s, CIA and Navy studies in Key West had been developing dolphin weapons systems. One of the few who acknowledged working on this project, James Fitzgerald, admits dolphins were trained for mine hunting and for attacking enemy divers, as well as surveillance.

In fact, the dolphin has become a preferred 'weapons system' of the navy. According to Tom La Puzza of San Diego's military Naval Ocean

Systems Center: 'The sonar system of a dolphin is superior to anything we can build. Until we can build a system as good as a dolphin's we'll keep using them.' By 1973, the navy was capturing over 100 dolphins a year for its marine warfare programmes.

In 1980, the *Washington Post* columnist, Jack Anderson, was leaked a top-secret Umbra CIA report claiming that the Russian dolphin programme with its five Black Sea research stations had reached such epic proportions that the agency feared a 'dolphin gap'. The Russians had gone so far as to develop 'anti-dolphin dolphins' which carried an 'acoustical jamming counter-measure to US Navy dolphin programmes'.

In 1976 former US Navy scientist Michael Greenwood revealed a range of military uses for dolphins. Greenwood testified that the navy had developed and implemented an extraordinary 'Killer-Dolphin' programme, later to be confirmed by Jack Anderson in the *Washington Post*.

The concept of the Killer-Dolphin became operational sometime between 1971 and 1973 in the waters of Vietnam's Cam-Ranh Bay, where the US fleet was anchored. There some navy dolphins searched for mines and worked with underwater demolition divers. Others served as sentinels to hunt saboteurs who swam into the bay at night. It is alleged that large hypodermic needles connected to high-pressure carbon dioxide cylinders were strapped to the beaks of dolphins which had been trained to attack enemy frogmen swimming near American ships in the waters of the Tonkin Gulf. Within a second of the needle's penetration, a massive injection of carbon dioxide was released with such force into the lungs or stomach of the victim that his body literally exploded.

The US military routinely refuses to acknowledge the existence of the killer dolphin programmes. However, when it was proved that hypodermic cartridges had been developed and mounted on dolphins for tests by the navy, a spokesman came up with the story that although the potential for using dolphins to kill people existed, the military had trained them to use their lethal needles on sharks, not people.

Not many swallowed this particular fishy story. Indeed, it is believed that up to sixty Vietnamese were killed by armed dolphins. Furthermore – although again the military refuses to acknowledge it – two American sailors were victims of killer dolphins during the Vietnam war. Evidently, the sailors had been drinking heavily while ashore, and upon returning to their ship one night, they fell overboard and were immediately blown to pieces by the innocent assassins.

More recently, when the mainly land-based Iran-Iraq War escalated into a sea war, the American navy was ready to respond with its ever-more-efficient dolphin squads. In 1987, six highly trained US Navy bottlenose dolphins and a twenty-man support team were flown into the Persian Gulf. It is believed this was just the first team of dolphins to be sent in. The navy has not commented on exactly what activities the dolphins are involved in, but it has indirectly acknowledged that they involve explosives and possible contact with enemy divers. Just how effective the dolphins have proved in this war is not known. It is known, however, that at least one of the original dolphins sent to the Gulf – a dolphin named 'Skippy' – has not survived the conflict.

9

Nuclear Whales

Trident missiles and retriever whales • Killer whales tracking submarines • FOE's Project Mobile Dick • 'Nuke the Whales' campaign.

Militarists are as prone to the 'bigger equals better' way of thinking as politicians and industrialists, so it was only a matter of time before the successful mobilization of the dolphin led to the partnership of the whale with the Trident nuclear submarine – probably this planet's most lethal weapon.

In May 1982, the Canadian branch of Greenpeace, as part of its ongoing protest against the Trident nuclear submarine base which serves the Pacific Fleet and is located just south of Vancouver, mounted a commando-style night raid to free two US Navy-trained white beluga whales that were on military exercises in Canadian waters. After the incident, Richard Meyer, a spokesman for the Naval Undersea Warfare Engineering Station in Washington State gave a detailed explanation of the whales' duties on the Trident and conventional submarine testing ranges in Washington's Hood Canal where the belugas had previously been stationed.

In military exercises, it emerged, dummy torpedoes are employed, but since these are extremely expensive whales are now being used to recover the test torpedoes. As Richard Meyer explained, sonar 'beepers' are placed in the torpedoes, so enabling the whales, with their own inborn sonar, to track the beeper and find the torpedo. Once located, the whale takes a harness clamp with long cables attached and slips it over the nose of the torpedo, after which it pushes a button and locks the clamp in place. The torpedo is then reeled in by the navy crews aboard the recovery ship.

The US Navy 'Project Deep Ops' which began in 1968 in Hawaii's Naval Undersea Center laboratories, and later continued with the Naval Special Warfare Group in San Diego, has trained numerous sea mammals for torpedo and equipment recovery at varying depths. For relatively shallow waters, California sealions and dolphins have been recruited; for depths of 850 feet, killer whales (orcas) are being used; and for depths down to 1,650 feet, pilot whales are employed.

'The programme is considered highly successful,' said Lt. Comdr. Mark Baker at the NSWG. 'The animals have proved extremely cost-effective.'

These revelations did not greatly surprise environmental groups. In April 1976, in the courtroom testimony of former US Navy scientist Michael Greenwood, it was disclosed that dolphins, trained by the navy and the CIA, had been instrumental in rescuing a nuclear bomb that had been accidentally dropped – fortunately undetonated – in the sea off Puerto Rico.

The use of whales to recover nuclear torpedoes was a logical progression from this and aroused the suspicions of environmentalists that whales, like the killer dolphins, were also being trained for more aggressive military jobs. They believe that at the very least the killer, beluga and pilot whales are being trained to plant limpet-mines and other explosives intended for enemy ships and most probably for some kind of anti-submarine warfare as well.

Their suspicions were confirmed to some degree by a 1983 statement by Marion Breland-Bailey, director of Animal Behavior Enterprises. Dr Breland-Bailey revealed that, under the auspices of the US Defense Department, she was researching the possibility of whales being deployed along the Russian coast to track and follow submarines so that they could be pinpointed by bombers.

Never missing an opportunity to ridicule the Pentagon's grand schemes, three members of the Sierra Club in California anticipated the next step in undersea warfare and offered the US administration a possible alternative to the extremely costly hide-and-seek MX missile system to be constructed in the Nevada and Utah desert. In an issue of the monthly Friends of the Earth newspaper, *Not Man Apart*, which, appropriately, appeared on April Fool's Day, they claimed to have devised a system that would be far cheaper and more devious than any previous inventions. The new system – an extension of the military's BBMs (Biological Basing Modes) would be called the MWX (Migratory Whale Experimental) or more commonly, Project Mobile (Moby) Dick. It would require the mounting of remote-controlled nuclear missiles on 200 of the 10,000 grey whales who, of their own accord, make the annual 13,000 mile journey from the Aleutian Islands to Mexico and back.

Project Mobile Dick is obviously superior to the land-based MX mechanized system, not least because the missiles have the advantage of being constantly on the move over vast areas and are therefore almost impossible to detect. It has also proved more reliable than some of the other BBMs which had failed in the experimental stages: the KPD (Kangaroo Pouch Developmental) the PBX (Pelican Bill Experimental) and the OWC (Ostrich Withdrawal Contemplative).

The authors rightly concluded that the MWX was a system whose time had come and Friends of the Earth issued car stickers throughout America to demonstrate its grassroots appeal. The sticker read: 'Nuke the Whales'.

10

Plankton Surveillance and Tracker Cockroaches

Submarine tracking plankton • Cockroach aphrodisiac used as means to trace agents • Other more sinister possibilities • 'The Combat Bedbug' and other CIA recruits.

Despite its 'bigger equals better' philosophy, the US Navy has utilized even the smallest forms of marine life as part of its military equipment. Even the plankton has been unknowingly enlisted as espionage agent and informer. If there is some doubt as to whether or not whales are being used to stalk submarines, there is no doubt at all that plankton is being so employed.

Plankton, like many creatures of the sea, pick up a good deal of phosphorescence and it seems that when submarines and warships pass through the massive shoals (sometimes hundreds of miles in size), it is easily possible, using satellite and laser technology, to identify their trailing tell-tale wakes – often some ten miles long – as a dark line cut through the phosphorescent surface.

Many animals, large and small, have also been employed as detectors. Canaries, originally used in coal-mines as a means of detecting gas leaks, were kept in the trenches in the First World War to give warning of gas attacks. In America today, technicians in nerve and mustard gas storage areas in the Rocky Mountain Arsenal north-east of Denver carry rabbits in wire cages to check cylinders in case of leaks; and when such gasses are transported, usually by rail, rabbits are kept as an early-warning system in each of the cars.

Professor Adolf-Henning Fucht, an East German scientist and later spy for the Allies, developed an airborne equivalent of the phosphorescent plankton informer, using fireflies. This was the Distant Early Warning Firefly. In laboratory conditons, the slightest trace of poison gas in the air could be detected by registering the electric light pulse of the fireflies.

In Göppingen, West Germany, goldfish are used as biological sensors to detect water pollution in much the same way as the fireflies detect air pollution. The goldfish emits constant electrical pulses which slow down when pollutant levels rise, thus alerting technicians to the rising pollution hazards.

Similarly, the Japanese use pet goldfish to give them advance warning of earthquakes. The fish pick up the advance vibrations and swim frantically in anticipation of the quake, thus giving their owners a few minutes to find a hopefully safe shelter.

In recent years the CIA has gained a reputation for involving itself in some very dubious operations and employing all kinds of low life in its investigations. Just how low this life has been, however, was not fully realized until the indefatigable Robert E. Lubow reported the results of experimental work undertaken on behalf of the CIA.

The 'Tracker Cockroach' concept was based on twenty-four years of research by a Dutch chemist who isolated the sex hormones of the female cockroach. The chemist used some 75,000 virgin female cockroaches in order to extract a minuscule two-hundred-millionths of a gramme of the hormone called periplanone-B. After analysing its chemical compounds, he synthesized it and found that one-hundredth of a gramme was sufficient to sexually excite 100 million male cockroaches. But whereas it was hoped through the original research to discover a cockroach repellant, the CIA's interest in the sex hormone of cockroaches, according to Lubow, was in order to use this as a tracking device, much like the infamous Russian 'spy dust'.

It was suggested that anyone sprayed or dabbed with this substance could be sniffed out by means of a small container of male cockroaches which had been wired for sound. Evidently, when sexually excited, cockroaches in common with fleas and many other insects emit highpitched squeals that can be picked up and amplified by sophisticated listening systems. Once marked by the scent, there is evidently no hiding from these bizarre agents.

Strangely, Lubow does not mention any other use for periplanone-B, but it would seem to have great potential as an instrument of psychological and physical torture. Anyone sprayed with this substance who was kept in a cockroach-infested area would have a very difficult time brushing off suitors. The idea of becoming the object of sexual desire for 100 million cockroaches is certainly a terrifying, skin-crawling one.

Appropriate as many people may feel this partnership of the cockroach and the CIA to be, it was not the only insect recruited by the agency. Using similar listening devices as those used for cockroaches, the General Electric Co., under military contract, set up an experimental station on Barro Island in the Panama Canal Zone. In a jungle environment similar to that of Vietnam, the human guinea pigs noticed that the various insects under observation – which included the giant nose-cone bug, the bedbug, the oriental rat flea, the mosquito and the tick – would sing and squeak at their approach. The best of these was the bedbug. Kept in a capsule with a built-in listening device, the 'Combat Bedbug' – as it was soon dubbed was considered the most successful at registering the approach of a human lunch.

7

The
Hospital
Corps

Strategies to save species from extinction

'What I have been preparing to say is this: in wildness is the preservation of the world." *Henry David Thoreau*

1

Noah's Children

Zoos of the past and future • Zoos as arks •
Captive breeding programmes: Gerald Durrell's
Jersey Trust • Pink pigeons and volcano rabbits •
Survival strategies of latter-day Noahs.

For the most part, zoos are destroyers of wildlife. Even today, large national zoos are often willing to buy critically endangered species to enhance the prestige of their collections, even though they are well aware that such species can only be obtained illegally. Smaller, private zoos often have even less scruples, and in many cases have proved to be fronts for black-market animal-trafficking operations.

It is no wonder that many of those concerned with endangered species are pressing for the closing-down of zoos. Gerald Durrell, who himself runs a kind of zoo on the island of Jersey, is one of these.

He is not alone in maintaining that, out of the tens of thousands of zoos worldwide, there are less than fifty with sufficient funds and expertise to provide adequate conditions for their animals. He also believes that a basic change in the philosophy and ethics of zoos is long overdue. His view is that while zoos may be a source of education and interest for humans, their essential concern should be the welfare and health of the animal species.

In Durrell's opinion, the only valid reason for taking animals into captivity is to run captive breeding programmes, or to study species in order to improve their health, or to save them from imminent destruction in their native habitat. His version of a proper zoo is something akin to Noah's Ark and, indeed, his own Jersey Trust 'zoo' could be compared with an ark. It is an institute primarily concerned with captive breeding programmes. Through the Jersey Trust, Durrell has achieved some remarkable victories. The pink pigeon of Mauritius, for instance, had fallen to a world population of only eighteen birds before he began a captive breeding operation which has rapidly revived the species both in captivity and in the wild. The Jersey Trust has similarly instituted successful breeding programmes for the volcano rabbit, various lemurs and marmosets, the brown-eared pheasant, the thick-billed parrot, the red-footed tortoise and the Jamaican boa.

If zoos have a future, many people now believe they must change from being wildlife consumers to wildlife producers. To their credit, some zoos like the Bronx, Basel and San Diego have gone some way to meet this challenge. Certainly zoo science has grown enormously in recent years in such areas as captive breeding.

Even the best zoos have their opponents. Many argue that preserving an animal's life in a zoo is a way of avoiding the problem in the wild. Furthermore, it allows the animal's habitat to be destroyed so that there is nowhere for it to return to, even if it is saved. In any event, environmentalists argue quite correctly, we just do not have the resources to protect all animals that are critically endangered by taking them into this kind of 'protective custody'.

The British organization, Zoo Check, is very clear on this point. Its director, William Travers, claims that zoos are squandering hundreds of millions of dollars on animal-preservation plans that would be better spent on the conservation of habitats: 'Despite a handful of captive breeding programmes undertaken by zoos, true conservation begins and ends in the wild. Once the habitat has disappeared, we will be faced with the preservation of a few species selected by man to share his "planetary cage" with him. As things are, zoos are just fiddling while Rome burns.' Zoo Check advocates a ban on taking animals from the wild, and an eventual phasing-out of all zoos in favour of natural habitat reserves or conservation centres.

Despite these arguments, it must be said that some zoos have played a part in the survival of certain nearly extinct species. Enlightened and obsessed eccentrics like Gerald Durrell have been the driving force behind many such rescue operations. It is certain that without the efforts of such individuals at least twice as many vertebrates would have met with extinction than has been the case during the last two or three hundred years.

Fortunately the number of organizations with mandates similar to that of the Jersey Trust has grown enormously during recent decades and these have joined the battle to save many critically endangered species from extinction. Some, like Durrell, work from individual motivation and private funding, others through universities and science institutes, and others still in connection with wildlife organizations, zoological societies or national government parks systems.

Whatever their affiliation, these men and women of mercy are real heroes in the ecology wars. Like Noah, they work hard each day, building arks for their beasts, all the while keeping an eye to the darkening sky and praying that their mission can be completed before the great flood comes.

2

Falcon Men of Ithaca

**Tom Cade and his amazing falcon-breeding hat •
Cornell University's Peregrine Fund repopulates
New England with several thousand falcons.**

Some may find it a peculiar way to make a living, but each day a worker firmly straps a strange black hat with a broad padded brim to his head and begins to do a little dance around the floor in a specially built loft room on the campus of Cornell University in Ithaca, New York. The shuffling dancestep does not seem particulary sophisticated, but the accompanying movements of the head with its strange hat are complex. The head is shaken, rattled, rocked and rolled in a distinctive and repetitive manner.

The dancer is not the only inhabitant of the loft room. High up, on a perch, is an elegant male peregrine falcon. Initially the bird seems uninterested in the dancer, but after a few moments the falcon begins to stare intently at the bobbing and weaving motions of the man in the hat. Suddenly it swoops from its perch and attacks the hat. Its claws clasp firmly as it strikes and, as it struggles and clings to the padded hat, the bird shakes its wide-spread wings to maintain balance.

It would probably take an uninformed observer more than a few moments to realize that he or she was not witnessing a predatory attack, but a carefully orchestrated mating ritual. The whole procedure is part of Cornell University's captive falcon breeding programme. The bizarre 'winged helmet' worn by the dancer is a device designed by falcon expert Lester Boyd, and is used to obtain sperm from the male peregrine falcon for the programme's artificial insemination scheme.

In 1970, Tom Cade, a professor of ornithology, established the Peregrine Fund as a long-term programme of research at Cornell University in order to learn how to propagate rare falcons in captivity. The programme was organized in response to the increasing rarity of peregrine and other falcons in the wild.

The peregrine is a bird celebrated for its hundred mile-an-hour hunting dives. Since ancient times it has been the falconer's favourite bird for its legendary speed and skill. By 1970, however, the peregrine falcon was clearly a vanishing species and ornithologists were despairing of its chances of survival. By that time, there were less than a hundred birds in the western United States most of whom were incapable of producing young, and the peregrine was already extinct in the east of the country. Similar declines were being observed in peregrine populations throughout its range in Scandinavia, Russia and Western Europe.

The reason for the disappearance of the peregrine was the wide use of pesticides. It took nearly twenty years for scientists to realize what was going on but, by the late sixties, it was shown that DDE, a breakdown component of DDT, interferes with calcium production. In predatory birds, like falcons, the presence of DDE in even the most minute quantities resulted in the production of thin-shelled eggs that could not withstand the weight of the nesting female bird. Consequently, no new chicks could be born to contaminated birds.

By the time DDT was finally banned in America, which was not until 1972, it seemed that it was already too late for the peregrine.

In North America, however, the American peregrine had two cousins that made their homes outside areas of heavy pesticide usage. These were the subspecies called the Arctic peregrine and Peale's peregrine. Unfortunately, the Arctic subspecies was also slowly building up critical levels of contamination on its annual southward migrations to South America. The only healthy subspecies seemed to be the non-migratory Peale's peregrine on the relatively remote Queen Charlotte Islands in British Columbia.

By now Tom Cade and a number of other falcon enthusiasts believed that they had nothing to lose by attempting a breeding programme with a few healthy captive peregrines. Most experts gave them little chance of success. Only three or four falconers had ever managed to breed the peregrine in captivity, and their efforts were widely considered one-off flukes which could not be seriously considered as a basis for a large-scale breeding programme.

However, in 1973, after three years of trying, Tom Cade's team succeeded in breeding twenty captive peregrines in a single year. Since then, the programme has been amazingly successful. Using a combination of traditional and unconventional approaches to the problems of falcon reproduction, they have bred hundreds of birds in their new, expanded facilities, and have released many back into the wild.

The bizarre falcon-mating hat was an innovation which proved particularly useful in producing sufficient quantities of healthy peregrine falcon sperm for artificial insemination. Evidently, one of the main obstacles to the production of fertile eggs in previous programmes was a chronic shortage of sperm, which was notoriously difficult to coax out of the average male peregrine.

With the advent of the mating hat, it was discovered that male falcons raised in captivity and 'imprinted' on humans could be sexually stimulated in a novel way. This was achieved by the wearer of the mating hat duplicating the ritual courtship dance of the female falcon. While it was found that the hat did not have to resemble a falcon to excite the male bird, the accuracy of the movement of the hat was critical. Once excited by the dance, the male bird would leap upon the hat in an attempt to mate with it. The hat was fitted with a heavy foam-rubber brim with a sperm reservoir or gutter in which the bird's semen could accumulate.

This strange device resulted in a hundred-fold increase in the acquisition of falcon sperm and has been a major reason for the increased reproductive rate in the Peregrine Fund programme. Thanks to Cade's inventiveness, peregrines have been returned to the wild and are now re-establishing themselves. Remarkably, they are regaining a foothold in the eastern United States where they had not been seen since the 1950s. This is the first time in ornithological history that a regional population has been revived by captive stock.

Not only are peregrines surviving their reintroduction to the wild, but they are now rearing healthy young birds. They have even proved that they are well-suited to high-rise living by making their eyries on the man-made cliffs of skyscrapers and office block towers in the great cities of eastern America.

3

Last of the Condors

**Captive breeding of many other rare birds of prey
• Debate over California condor • Interventionists'
case • All wild condors captured.**

Since Cornell University's captive-breeding programme of the early 1970s a large number of similar projects involving birds of prey have been initiated throughout the world. Besides peregrines, there have been successful captive-breeding programmes for over fifteen falcon species, including gyrfalcons, sakers and lanners. Other spectacular birds of prey that have been granted an eleventh-hour reprieve are the red-tailed hawk, Harris's hawk, the ferruginous hawk, the osprey, bateleur, burrowing owls, the common buzzard, the lesser spotted eagle, the Philippine monkey-eating eagle, the bald eagle and the golden eagle.

Perhaps the most controversial of all captive-breeding rescue plans has been the attempt in the 1980s to save the California condor. With a wingspan of over nine-and-a-half feet, the condor is America's largest bird, and certainly one of its rarest. The condor has had protected status for longer than any other bird of prey in America, yet it has continued to be persecuted, and the threat of extinction is greater now than ever before.

For most of the century, the California condor population has remained at between forty and fifty birds. By 1980, it had diminished to less than thirty, with no sign that any population increase was possible. Reproduction in the condor is slow and precarious. Each pair of birds seem to lay only one egg every two years on bare earth, often on rocky ledges, and the egg takes fifty days to incubate. The art of flying takes a considerable time for young condors to learn, and the young bird will remain for nearly two years with its parents. The condor takes six years to develop adult plumage, and at least ten years to reach sexual maturity. The only factor working in its favour would seem to be its comparative longevity of about fifty years.

Considerable debate was raised over the captive breeding of the condor. Those in favour of it proposed that all remaining condors be captured, in order to implement their rescue plan; those against argued that the risk of failure was too high, and that, even if the programme was successful, it would remain to be seen if captive-reared birds could be reintroduced to the wild.

The case of the interventionists was not helped, in 1980, when one researcher examined and handled a chick for over an hour despite its obvious distress. The incident resulted in its death by shock. Thus, those who are attempting to save the species ended up killing one of the few fledglings that had managed to hatch in a decade. On the other hand, interventionists pointed out that other predatory birds were responsible for destroying the occasional egg. To further support their argument, they produced a film showing an egg being accidently knocked off its rocky ledge by the squabbling condor parents, themselves.

Some naturalists maintained that even if the species was on its last legs, it should be left alone. If it could not be saved in the wild, it should at least be allowed to drift into extinction with dignity and freedom. Its last days should not be spent as a caged and captive creature.

The captive-breeding lobby finally won the day, helped by the fact that captive breeding of the California condor's nearest relative, the Andean condor, had recently been achieved successfully. Encouraged by new breeding methods, captive Andean condors had increased their population rate by up to 500%.

By the mid-eighties, the extremely difficult task of safely trapping the Californian condors had been carried out. The last 27 Californian condors left on earth were taken into 'protective custody'. In 1988, the programme achieved its first success with the hatching of a captive-bred California condor. Feeding problems inevitably arose and these were ingeniously overcome by the young bird being fed its meal of minced mouse by a handpuppet resembling the beaked head of a mother condor.

The fate of this magnificent bird now lies entirely in the hands of those who run the captive-breeding programme. It will be a very long time before we know whether the survival of the California condor is a possibility. And even if the programme proves successful, it will be even longer before we know whether this great survivor of the Ice Ages will ever be able to return safely to its wilderness homeland.

4

Whooping It Up

Whooping Cranes • Population of 15 in 1941 •
Protection and 'double-clutching' programme •
Sandhill cranes as foster parents to whoopers •
International Crane Foundation • The odd couple:
Archie and Tex • Rescuing the cranes of the world.

'If you've seen fifteen whooping cranes, you've seen them all.' In 1941, this statement was the literal truth. By 1962, after two decades of total protection, the world population of these tall, elegant creatures, with their distinctive red and black markings, had only risen to twenty-eight. Although this was an improvement, the fact remained that the crane species was limited to a single flock of birds that twice a year ran the gauntlet of a 2,600 mile migration route from northern Canada to southern Texas. All that was needed was one fierce storm, a serious pollution incident, or even an enthusiastic gang of hunters to obliterate the species. Beyond protection from the hunters who had pushed them to the edge of extinction, and of their summer and winter territories, something needed to be done to boost the reproduction rate of cranes.

In 1967, the Canadian Wildlife Service started a programme of nest-robbing. It was a strategy similar to the 'double-clutching' system used in the falcon captive-breeding programme. However, it had been discovered that, unlike falcons, whooping cranes did not lay again if their eggs were stolen; they usually laid two eggs and almost without exception only one of the chicks survived. With this fact in mind, wildlife wardens made helicopter raids on nesting sites in Canada's Wood Buffalo National Park. The raiders took one egg from each nest and stored them in special heated pouches. The eggs were then loaded on to a jet aircraft and flown to the Patuxent Wildlife Research Center in Laurel, Maryland, where a special breeding facility had been set up.

In the first six years of the programme, sixty eggs were collected and over twenty birds had survived. Furthermore, in that time, two of the captive birds had themselves laid and hatched an egg. In the same year, 1975, an extremely novel foster-parent programme was put into operation, in which some of the 'surplus' whooping crane eggs were placed in the nests of their nearest relatives, the slightly smaller but more numerous sandhill cranes.

Much to the surprise of many ornithologists, the experiment has been extremely successful. Despite the obvious difference in size and colour, sandhill parents have reared and protected the whoopers as their own. Even more importantly, the fostering programme seems to be producing crane populations with different migration patterns to the main flock. The eventual plan is to establish several populations of whooping cranes with different nesting and wintering grounds and different migration patterns to safeguard the species against the possibility of extinction through a single natural or man-made disaster.

By 1977, only ten years after the egg-collection policy had been instigated, a record number of 126 whooping cranes was counted. It was the largest crane population in existence in over sixty years. Moreover, it was a seemingly healthy population with many surviving young birds, each looking forward to a full twenty-five years of life. Naturalists were beginning to feel hopeful that the whooping crane would be re-established in considerable numbers by the middle of the next century.

That same year, 1977, the International Crane Foundation was established in Baraboo, Wisconsin. Working in association with the Patuxent Center, the founders of ICF, George Archibald and Ron Sauer, seized the public imagination by their innovative methods of inducing artificially inseminated cranes to lay eggs with the stated aim of 'saving the cranes of the world'.

Both men were Cornell University graduates – Archibald's degree was, appropriately enough, in crane behaviour – and, doubtless inspired by Tom Cade's falcon experiments. However, the difficulty with cranes was not so much in obtaining sperm from males or in inseminating captive females, but rather that females must perform their ritual mating dances with a man in order to become sufficiently excited to produce an egg. This is because, having been reared in captivity, female birds are not interested in male cranes; they have been 'imprinted' on humans rather than other cranes.

One such bird was a female crane called Tex who had been raised as a chick in the San Antonio zoo in 1967. She was taken to Patuxent ten years later in the hope that she would breed. It was immediately evident that Tex thought of herself as more of a person than a crane. When Archibald visited the centre, he found that she seemed to be especially attracted to him and he soon took advantage of it. In the manner of the falcon experiments, each spring, for four years, the 'odd couple' performed the ritual whooping-crane mating dances that Tex had refused to perform with other whooping cranes but was more than happy to do with Archibald.

It was not until 1981 that these intensive pre-natal aerobics paid off. Having artificially inseminated Tex, Archibald lived with her for six weeks, for four of these he danced with her several times a day. He also foraged for twigs to help Tex build her nest. Finally Tex laid an egg and two-and-a-half weeks later the egg hatched. The little whooper survived its unorthodox upbringing and was named Gee Whiz.

For Archibald, his colleague, Sauer, and the International Crane Foundation, whooping cranes are just one of many species in serious trouble. The Foundation, by paying close attention to crane behaviour patterns, has achieved considerable success in its captive-breeding programmes. By 1980, ICF was already captive-rearing twenty-five rare cranes a year, and has now captive-bred specimens for all fifteen of the world's crane species, including the rare Japanese red-crowned crane and the even rarer Siberian crane.

Despite his remarkable success in the breeding of captive cranes, Archibald recognized such programmes for what they are: first-aid measures to revive dying species. He also realizes that the acquisition and maintenance of a suitable natural habitat is the only long-term guarantee that a species will be saved for future generations. It is for this reason that Archibald considers ICF's greatest success to date to be the establishment of a unique wild habitat for cranes. The Foundation played a crucial role in getting the Han River estuary – in the uninhabited Demilitarized Zone which separates North and South Korea – declared a forty-square-kilometre marsh park for rare cranes and other endangered species.

5

Search and Rescue

Successful captive breeding of wild hoofed mammals • American and European bison • Père David's deer, Przewalski's wild horse.

Some of the most spectacular recoveries of nearly extinct animal species have been achieved by individuals who have orchestrated extensive 'search-and-rescue' operations to save some of the wild hoofed mammals. Success in this area of the animal kingdom is perhaps not all that surprising. Through the domestication of cattle, goats, sheep and horses, the human race has accumulated considerable knowledge of the reproductive processes of hoofed animal species.

In America, the famous Great Plains bison or buffalo herds once boasted a population in excess of 60,000,000. By the end of the last century, market hunters had virtually wiped them out. The eastern black bison subspecies was extinct, as was the far western Oregon bison, while the bison of the Great Plains were reduced to less than a hundred animals roaming the wild.

In 1889 the government policy of intentional extermination (in order to starve out unpacified Indian tribes) was reversed, and a search-and-rescue operation was launched. America's last eighty-nine free-roaming wild bison were captured that year and put on protective reserves. A further few hundred bison had survived for some time behind the protective fences of Yellowstone National Park and a national buffalo park in Canada. Unfortunately, for the most part, these captive herds were either not in good health or were beyond breeding age. Also, this first rescue plan had been poorly funded, with the result that the animals were expected to forage for themselves. On their limited reserves it was feared that a particularly severe winter or a single epidemic might result in the herds' extinction.

The fact that this did not come about was largely due to the considerable efforts of William T. Hornaday of the New York Zoological Park. It was Hornaday who founded the American Bison

Society and raised large amounts of money to promote breeding programmes and allow these stranded herds, deprived of freedom to migrate to warmer southern grasslands during the winter, to be fed and sheltered. The Society reached a turning-point in 1905 when President Theodore Roosevelt threw his support behind it. Luckily, bison have proved to be strong breeders. From these core herds, the Great Plains bison made a stunning recovery. By 1912 there were approximately 3,000 animals in existence. Today there are close to 40,000 bison in North American wilderness parks and it is now clear that, so long as their parklands remain secure, the North American bison have been saved from extinction.

An even narrower escape was achieved by the forest-dwelling European bison or wisent.

In 1923, by which time both subspecies of the wisent had become extinct in the wild, the Polish zoologist Jan Sztoloman founded the European Bison Society at the Berlin Zoo. Without the action of this group it is likely that the European bison would have entirely vanished. In 1925, an ageing bull called Kaukasus died in Hamburg Zoo, and with its death, the Caucasian wisent was truly extinct, for it proved to be the last captive animal of its race.

Sztoloman's Bison Society was more fortunate with the northern race of wisent. It managed to acquire a total of six captive animals from three different collections to form the nucleus of a breeding group. Today between two or three thousand European bison roam woodland parks and reserves in Europe, the largest herd in the wild being in the Bialowieska Forest in Poland.

Two other extraordinary stories of survival concern a unique form of deer – Père David's deer – and the last species of undomesticated horse – Przewalski's wild horse. Neither animal would have avoided being hunted to extinction if it had not been for the efforts of a few enlightened men of influence.

.Père David's deer had a particularly narrow escape. It was first identified in 1865 by Father Armand David, the French naturalist-priest who discovered the giant panda. It was a large and unique 'horse-tailed' Asian deer which had evidently become extinct in the wild some time before 1800.

When Père David first saw the deer, the only remaining population survived in a walled Imperial reserve. Tragically, these were slaughtered by European troops in Peking during the Boxer Rebellion in 1900.

By great good fortune, the Duke of Bedford – an enlightened private collector and passionate amateur zoologist – decided to acquire some of these unique animals. The eighteen Père David's deer at Woburn Abbey proved to be the salvation of the species. They bred well in the open country of the deerpark at Woburn and today there are over a thousand Père David's deer in more than a hundred parks and zoos around the world. The largest herd still remains at Woburn Abbey, and recently a few of these animals have been returned to the old imperial parks in China.

Przewalski's wild horse is the only undomesticated horse species to survive into the twentieth century. It is a short, stocky animal with a gold-brown coat, a black mohawk-style mane, and a long black tail. It was discovered in 1879 in the Gobi Desert by the Russian scholar, Nikolai Przewalski. At that time there was one other true wild horse: the tarpan, which lived in the steppeland of southern Russia. These two animals were the true ancestors of all domestic horses. However, the grey-coated and rugged little tarpan was so ruthlessly hunted that, by 1887, it was extinct, and Przewalski's wild horse was the only Ice Age horse left in the world.

Przewalski's wild horse now appears to be extinct in the wild. However, because of Przewalski's discovery and the acquisition of some fifty animals between 1897 and 1902 by Western collectors (including the beneficent Duke of Bedford) and zoos, captive-bred animals have survived.

In 1956, there were only 36 of these, but due to careful and improved breeding techiniques (particularly in the two main captive herds in Prague and the Catskill Mountains in the United States) their numbers had increased to 250 by 1976. Today there are more than 500 of these unique horses, and plans are afoot to restore some to their Mongolian homeland to form the basis of a new free-roaming herd.

6

Operation Oryx

Arabian massacre of 1961 • 1962 rescue plan for
last of the Arabian oryx • Programme succeeds •
Species is reintroduced into the wild.

One of the most dramatic and widely publicized of all search-and-
rescue operations launched to save a mammal species from extinction
was that of the Arabian oryx. The oryx is often considered the most
beautiful of all the world's antelopes. It is a large, cream-coloured
desert antelope with magnificent, two-and-a-half-foot-long spiralling
horns. Once it roamed throughout the deserts of the Arabic world.
However, by the beginning of this century, relentless hunting of the
Arabian oryx had resulted in disappearance everywhere except the
Rub al-Khali, the 'empty quarter' of the great southern desert of
Arabia, Aden and Oman. By the 1950s, even these remote surviving
oryx herds came under extreme pressure. Large motorized hunting
parties, organized by wealthy Arabs, found it amusing to use
automatic machine-guns to shoot at the fleeing oryx from speeding
jeeps, or even airplanes and helicopters.

In 1961, by which time it was estimated that there were less than a
hundred wild Arabian oryx left in the world, a party of hunters from
Qatar on the Persian Gulf entered Eastern Aden and over several
weeks, tracked down and killed no fewer than forty-eight animals. It
was after this massacre that conservationists decided a major search-
and-rescue operation must be put in hand. Thus, in 1962, the Fauna
and Flora Preservation Society of London and the World Wildlife Fund
(WWF) launched 'Operation Oryx' in an attempt to save the species.

An expedition headed by Major Ian Grimwood was sent into the Rub
al-Khali desert to track down and capture a herd of oryx so that a
projected captive-breeding programme could be established. The
expedition revealed that the plight of the oryx was far more serious
than even the most pessimistic estimates had suggested. In six
thousand miles of desert, the searchers were only able to locate four
oryxes. With considerable difficulty these animals were captured alive

– although one died shortly afterwards from a bullet which had remained in its body from a previous encounter with hunters.

The three surviving wild oryx were placed in a special breeding unit of Phoenix Zoo in Arizona. There they were joined by one captive animal from the London Zoo, one from the Sultan of Kuwait, and four from the private collection of King Saud. These nine animals, which were named 'the World Herd', were the nucleus of the captive-breeding operation. Within fifteen years, the captive population grew to over a hundred healthy animals, and breeders were sure that the critical period was over.

Meanwhile, in the Arabian desert, the 'empty quarter' became emptier still. Operation Oryx was indeed an eleventh-hour rescue plan, for the animal became extinct in the wild soon after. The last three Arabian oryx to be seen in the wild were shot in 1972. Nor was the oryx the only victim of these senseless hunters; most other forms of large game had also been eliminated. In 1966, the last Arabian ostrich ever seen was shot, and as there were none in captivity the race was now totally extinct.

Luckily, by the 1980s the Arabian oryx had bred so successfully that the decision was made to reintroduce captive animals into certain protected wild regions. To this end, release sites were chosen in Israel, Jordan and the wilderness of Jiddat-al-Harasis in Oman. The governments of each nation have set aside land and have introduced reasonable security measures. In Oman, for instance, the local Harasis tribesmen are effectively employed as rangers to guard the herd from poachers.

The captive oryx have readily adapted to the wild, and the possibility of large oryx herds once again roaming the Arabian deserts is now far more of a practical probability than a conservationist's dream.

7

Save the Tiger and the Hamster

World Wildlife Fund's tiger rescue programmes •
Three races extinct, four endangered • Success in
India • Rescue of golden hamster • Population
rises from 13 to millions in 25 years.

Like Operation Oryx, the Save the Tiger appeal, a decade later, was a
large-scale, high-profile rescue plan organized by the World Wildlife
Fund, and largely inspired by the efforts of Guy Mountford. Launched
in 1972, it was the WWF's largest project up to that time and required
careful negotiations in order to win the co-operation of the
governments of several countries.

As with Operation Oryx, it was recognized that extensive fund-
raising would be necessary if the project was to achieve any long-term
effect. The WWF proved equal to the task of raising several million
dollars in donations, as well as achieving comparable grants from the
various governments. However, the organizers were aware that time
was running out. The tiger population was spiralling rapidly
downwards in all its territories.

In common with all large predators, the tigers were being pushed
towards extinction by a combination of the destruction of their habitat
and the activities of black-market hunters. This century alone has seen
the extinction of three types of bear, two types of lion, one jaguar and
over a dozen wolf-like predators – from the miniature Japanese wolf to
the strange, carnivorous marsupial called the thylacine from
Tasmania.

Of the eight subspecies of tiger, by 1972, three – the Bali tiger, the
Javan tiger and the Caspian tiger – were extinct and four species were
critically endangered. The Siberian and Chinese tigers were down to a
couple of hundred and data on Sumatra and Indonesian subspecies
was incomplete. Only in India was there a considerable population of
two thousand or so tigers, but even this was dwindling quickly.

Unlike many endangered species, breeding was not really the problem. Tigers were productive enough. The problem was habitat. Tigers needed land in order to build up their populations and they needed protection from poachers once reserves were established.

Logically enough, the greater part of the rescue effort was directed towards the Indian subcontinent, where the chances of success seemed most likely. And indeed they were; the project workers were rewarded by remarkable results in a relatively short time. With essential local and national government involvement, a score of large reserves were established covering over eight thousand square miles, and the Indian tiger population has doubled to about four thousand.

Elsewhere, encouragement and aid from the WWF's Save the Tiger initiative has heightened international awareness of the problem. The population of the Siberian tiger has increased to 500, and Chinese populations to 250 or more. The Sumatra population now appears to be around 500, while the Indo-Chinese race may be as large as 2,000, despite extensive and ongoing poaching operations.

Most, though not all, last-minute search-and-rescue operations need to be followed up with long and arduous population rebuilding programmes. In a few cases, once a species has been saved from the immediate threat of extinction, it has displayed a remarkable ability to recover and regenerate.

One of the greatest zoological comebacks of all time occurred in 1930, when a Palestinian scholar named Professor Aharoni read in ancient Aramaic texts about a curious little animal that was popular with children as a domestic pet in times of antiquity.

This 'special kind of Syrian mouse, which was brought to Assyria and into the land of the Hittites' was unlike any species the professor had ever come across. His curiosity was sufficiently aroused for him to embark on a personal quest to the ancient ruined Hittite city of Chaleb, where the texts claimed that the mouse was common.

After a diligent search of the region of what was once Chaleb (and is now Aleppo), the professor discovered a burrow in which were entrenched a group of thirteen of these unique red-gold rodents. What Professor Aharoni had discovered was the world's last surviving population of the now common and popular household pet called the golden hamster. No one in modern times had ever recorded sighting these animals before and, despite extensive searches, they have never again been found in the wild.

Extraordinary as it may seem, the world's entire population of pet hamsters – which now runs into tens of millions – have all been bred from those last thirteen specimens recovered by Professor Aharoni in 1930.

8

Saviour of
the Dodo Tree

The riddle of the dodo and the tree • Dr Temple's turkey experiment • Mystery solved • Other plant extinctions.

Dr Stanley Temple did not set out to become a green-thumb saint; his real interest was in birds. In fact, he was a raptor specialist who had been – with Tom Cade – one of the initiators of the Cornell University peregrine falcon captive-breeding programme. Something like the momentum of Fate drew him to Mauritius.

In the annals of ornithology, Mauritius will go down as the site of more bird extinction per square mile than any other part of the planet. The Mascarenes Islands – a group of three small Indian Ocean islands off the coast of Madagascar, of which Mauritius is one – can collectively claim no fewer than thirty extinct animals, twenty of which were unique bird species.

These have included several colourful and unusual species of parrots, parakeets, owls, pigeons, and rails – not to mention the indigenous lizard, snake and various giant tortoise species. However, Mauritius is undoubtedly best-known for that huge, fifty-pound flightless dove called the dodo. It was here that the last of the dodos – and his closest relative, the white dodo, and two more elegant cousins, the swan-sized birds called solitaires – were hunted to extinction about three hundred years ago.

On visiting Mauritius in 1973, Dr Temple became curious about the seemingly symbiotic relationships between certain plant and animal species. In Hawaii, for instance, the extinction of ten species and twenty subspecies of the nectar-feeding tropical birds called honeycreepers and honeyeaters had resulted in many flowering trees following them into extinction. The reason for this is fairly obvious. These birds had naturally evolved symbiotically with the trees and served as the means by which their flowers were pollinated. Once the birds vanished, the trees eventually disappeared too.

On Mauritius, however, Dr Temple soon became aware of a more mysterious symbiotic relationship between plant and animal species. This involved the extinct dodo and the tree Calvaria major which is endemic to Mauritius and survives as a species in only thirteen living specimens. Its fruit was known to provide a major source of food for the dodo, hence the Calvaria's local name, the 'dodo tree'. But the relationship between bird and tree appeared to go further than that. Historic records show that the dodo tree was common on the island when Europeans first arrived and yet there was no evidence that a single new tree had appeared since the extinction of the dodo in 1680.

Obviously, some kind of symbiosis had existed, and the ability of the tree to regenerate was lost when the dodo became extinct. The strange thing was that the few remaining 300-year-old trees were producing seemingly healthy fruit and fertile seeds, yet despite many attempts to grow them, the seeds failed even to germinate. With all thirteen remaining trees nearing the end of their lifespan, the situation was becoming critical.

Dr Temple was determined to solve this intriguing riddle. He began with a series of experiments in which he force-fed a number of turkeys with dodo-tree fruit. These experiments led him to the conclusion that dodo-tree seeds would only germinate after they had been ingested by a dodo or similar large bird. It was necessary for the thick seed-casing to be scoured or crushed by the bird's gizzard. Later, when the bird's droppings were scattered over the forest floor, the seed germinated and the seedling appeared.

By collecting the droppings of the turkeys, Dr Temple managed to germinate three seeds – the first for three hundred years. It is now thought that some system of artificial abrasion of the dodo tree's seeds can readily prepare it for planting. By this means, it is now firmly believed, the species can be saved from extinction.

From the case of the dodo tree, it will be seen that last-minute search-and-rescue operations have not been limited to endangered bird and animal species. Indeed, the dodo tree is only one among thousands of rare or endangered plant species to have been rescued. Today, many scientists argue that an even more critical, if less publicized, ecological crisis exists in the plant world than in the animal world. Scientists are grimly aware of how important it is to conserve the vast genetic resource library that the world's plant life represents, and many seed collections and protected gardens have now been established.

Since the extinction of the dodo, there have been approximately three hundred vertebrate animal extinctions. In the same time span it has been estimated that there have been perhaps ten thousand plant extinctions. It takes very little knowledge of natural history and ecology to understand the survival of the earth's plant life is crucial to the survival of all other forms of life on the planet.

9

The Frozen Zoo

Captive breeding experiments • Hybrid species: tiglons, beefalo, ratmice, siabons • 'Reconstructing' extinct species • Surrogate mothers; interspecies embryo transfers • San Diego's Frozen Zoo sperm and ovum bank • Mammoth recreation experiment • The future: genetic engineering and cloning.

Zoos and captive-breeding programmes have bred more than a few chimeras. There are a considerable number of tiglons and ligers – tiger and lion hybrids – in captivity. There are large herds of commercially raised beefalo – a cross between a bison or buffalo and domestic cattle – on a number of American ranches. There are colonies of ratmice, and there is a monkey called a siabon or gibbang – a cross between a gibbon and siamang.

A couple of programmes have taken traditional selective breeding techniques and turned them on their heads. Such programmes were instituted in recent years in an attempt to 'reconstruct' the extinct wild bull, the aurochs, and the extinct wild horse, the tarpan. The results have been the subject of some debate. It would certainly be difficult to argue that the products of these programmes are the same as the wild species raised Lazarus-like out of the oblivion of extinction. What in fact has been achieved are domestic animals with most of the outward physical characteristics of the lost wild species.

Although seemingly a logical approach to achieving the revival of these animals, there is something reminiscent of science fiction in the idea of breeding an animal so that its own offspring become its ancestors. It is rather like the plot of a time-travelling novel.

As mentioned earlier, many rare birds and animal species have been successfuly bred through a combination of artificial insemination and fostering methods. However, the science of such programmes has now gone far beyond these concepts into the even more sophisticated area of surrogate motherhood and interspecies embryo transfer.

In 1977, a mouflon – a wild Sardinian sheep – was born from a domestic sheep at Utah State University. In 1981, a gaur – a wild Himalayan ox – was born in the Bronx Zoo from a domestic Holstein cow named Flossie. These were the first two successful experiments of their kind. The results were not crossbred animals, but purebred wild species carried by domestic surrogate mothers. The process involved acquiring a fertilized embryo from a wild female and placing it in the womb of a domestic host mother.

In the case of rare or endangered wild species, the advantage is that the rate of breeding may be increased dramatically. To take the gaur as an example: instead of a natural rate of a maximum of one gaur calf a year, a pair of gaur and a stable of surrogate domestic cows could produce between six and eight calves a year.

Preparations for this 'brave new world' of embryo transfer and surrogate motherhood are already under way. Since the late 1970s, the San Diego Zoo has been building up an alternative 'Frozen Zoo', in which fibroblast cells taken from the connective tissue of hundreds of endangered species are stored in plastic vials of liquid nitrogen. This sperm and ovum bank is a kind of insurance policy. It is hoped that the absolute cold of liquid nitrogen (-196°C) is enough to preserve cells and fertilized eggs and tissue until a future technology is sufficiently developed to recreate the animals if they become extinct.

Just how strange this new area of science is becoming can perhaps be gauged by the bizarreness of some of the experiments that are currently in progress. In 1980, the Soviet Academy of Science reported that it was confident it coud recreate a living specimen of the woolly mammoth. The Soviet proposal was to reconstruct an Ice Age mammoth by thawing frozen mammoth tissue taken from glaciers and implanting the nucleus of cells from this tissue into the fertilized egg of an elephant. The egg would then be placed in a female elephant's womb, and the host mother would give birth to a prehistoric woolly mammoth.

Difficulties inevitably arose, however. With the limitations of current technology, the Russians were unable to revive the long-frozen cells of the ancient mammoth and this led to the collapse of the experiment.

Genetic engineering is an extraordinary new field with almost limitless possibilities. However, it is ironic that we are rapidly losing vast genetic reservoirs in plant and animal extinction at just the point when we are developing the technology to make use of them.

Furthermore, although we are seeing extraordinary advances in breeding techniques, such as embryo transfer, cloning and genetic engineering, it must be recognized that, so far as endangered species are concerned, these are desperate last-minute tactics at resurrecting species. The fact remains that it is far easier and cheaper to 'save' a species before its condition becomes critical. For most species, the serious implementation of a few basic conservation policies would suffice. Without in any way wishing to undermine the – often astonishing – achievements of the men and women in the 'Hospital Corps' who have literally brought species back from the brink of extinction, it must be said that the best and surest way to save a species is to make sure that it remains safe and secure in its own natural habitat.

8

Biological Weapons

History and future of biological weapons

'Soldiers have rarely won wars. They more often mop up after the barrage of epidemics.' *Hans Zinsser*

1

Beetlemania

A theological debate: Haldane's view of God as
an entomologist • Evolutionary and numerical
superiority of insects • Homo sapiens' place in
evolution.

In an historic BBC radio programme, a startling juxtaposition of
religion and science came about in a debate between the Archbishop of
Canterbury and JBS Haldane, one of the founding scientists of
evolutionary biology.

A poignant moment in the debate was reached when the Archbishop
put a theological question to the scientist. 'Given your lifetime study of
the Creation, what would you conclude as to the nature of the
Creator?' The scientist's immediate response must have somewhat
bemused the Archbishop. 'An inordinate fondness for beetles,' replied
Haldane briskly.

Haldane's statement may seem whimsical to the layman, but it is
based on solid scientific observation. If we release ourselves from our
narcissistic, human-centred view of nature and look at the earth
objectively, the inevitable conclusion we reach is that the planet is
dominated by insects.

Considering the various forms of life that inhabit the earth, we find
that the vertebrate species of birds, mammals and reptiles comprise
less than 20,000. If we examine the plant kingdom, we can catalogue
approximately 300,000 species. However, when we come to investigate
the insect species we discover that there are well over a million of
these. Furthermore, of all of these insects, one-third are beetles.

Focusing on this planet of insects, it is interesting to observe that
there are nearly as many species of cockroaches alone as there are of
mammals. Numerically, insects outnumber humans by twelve million
to one, and, even given our hugely superior weight, insects outweigh
us by approximately four to one in total biomass. Further, we know
that insects have resided on this plant for at least 370 million years, a

much longer timespan than the mere two million years of human occupancy.

Based on such data, what conclusions might an objective observer reach about the nature and concerns of the Creator? It is a curious fact that the ancient Egyptians chose the image of the scarab beetle as their most common decorative motif. Sometimes they carved the beetle's image in solid limestone in the immense size and proportions of an army tank. Did the dominance of the beetle motif betray an ancient knowledge of the true order of life forms that we are now just rediscovering?

When man looks at the natural world and the threats to his primacy in the chain of being, he usually thinks of huge man-eating predators as his most formidable adversaries. In fact, his most dangerous opponents are the smallest ones. The mosquito alone kills more than a million people each year. The common housefly spreads enough disease to kill half that number, while the oriental rat flea, carrier of bubonic plague, has exceeded all records for genocide, wiping out as many as 25 million people in a single epidemic.

Throughout history, untold millions have died of famine brought on by the onslaught of the Biblical desert locusts (a short-horned grasshopper) which, in vast clouds of as many as 25 billion insects, have the capacity to cover 2,000 square miles and consume 3,000 tons of food crops in a day.

And although not in the same league as the mega-killers of the insect world, it is a startling fact that bees and wasps are responsible for more human deaths each year than all the sharks, crocodiles and other large 'man-eating' predators combined.

2

God as Microbe

Bacteria as the dominant life form on earth • The
microbe's perspective • Malaria and other killer
microbes • H. G. Wells' *War of the Worlds*.

Formidable and numerous as insects are, however, if we consider all
life forms on this planet, we must conclude that Haldane's perspective
was not quite radical enough. While insects are without doubt the
most numerous and successful *visible* creatures on the planet, of the
estimated 3,000 quintillion (that is, 3,000,000,000,000,000,000,000,-
000,000,000,000) living things, 75% are bacteria. These unicellular
micro-organisms, the minutest and earliest forms of life, can survive
the most extreme conditions of heat and cold. They can be found in the
deepest ocean and the furthest stratosphere. Even in the event of
nuclear explosion, they have survived a radiation dosage 10,000 times
stronger than that which is fatal to humans.

If microbes had minds at all, there would be little doubt in those
minds as to who the Creator had made this planet for. Indeed, so long
as life remains on the planet, microbes will make up the vast majority
of it. Further, microbes always have and always will dictate the
conditions by which all other life-forms are permitted to live.

It might be instructive to view the war between man and his most
dangerous visible enemy in the animal world, the mosquito, from the
perspective of a microbe.

There are some three thousand species of mosquitoes. Most are
harmless to humans, but a few are deadly. The most dangerous is the
anopheles or malarial mosquito. This insect is the real Dracula of the
animal world. Its nocturnal bloodsucking habits keep six hundred
million people a year permanently enthralled with shivering fevers and
crippling illness. Each year it claims the lives of one million of these
fevered victims. But the real killer is not the insect at all. The sting of
the mosquito would be no more than a minor irritation were it not for
the invisible demon cargo that is carried and reproduced in the belly of
the female mosquito. The malaria microbe is a being that attacks the

actual life-blood of the human species. It invades the red corpuscles and destroys them.

A somewhat bleak view of the 'nature of the Creator' might conceivably emerge from considering the life-cycle of the malarial microbe and man's role in it. Man, his suffering, and even his death, are of no consequence to the microbe. From its point of view, the only purpose of malarial fever is that it makes the human an attractive hot snack to a passing anopheles mosquito.

The cycle of malaria results in a victim's fever 'breaking' at dusk each day. The victim bursts into a heavy sweat that is a climax to the day's suffering and finally allows him to fall into an exhausted sleep. This is the perfect lure for the mosquito – an aromatic dinner bell for the smell-oriented insect at dusk, the peak-time for mosquito activity. Further, the victim's exhausted sleep leaves him unresisting and undefended, making it all the easier for the mosquito to feed on the poisoned blood.

And this is the whole reason for man's suffering. Six hundred million humans a year sicken and suffer in order to allow this microbe to travel from the human blood cells into the safe harbour of the belly of the female anopheles mosquito. For here alone, the microbe may enter the adult sexual stage of its existence, and from here alone it may send out its many monstrous children to destroy the lives of millions of suffering humans.

So, what we may view as the high tragedy of human suffering and death from this perspective is nothing more than a bed-and-breakfast service for microbes travelling from one mosquito hotel to the next.

Microbes are the most formidable, ruthless and effective of killers. Adaptable and wily beyond all cerebral intelligence, they are genetic generals quite without compassion.   Malaria kills one million people a year, and so do polio, diphtheria, tuberculosis, whooping cough, and measles. The diarrhoea microbe alone kills five million. Fortunately, the killer microbes are a tiny minority, or we would never survive. Most are sympathetic to our continued existence; for humans, along with most other animal species, are a useful means by which microbes prosper, breed and multiply.

H. G. Wells revealed the identity of the real lords of life on this planet in his famous novel, *War of the Worlds*. When alien beings of infinitely greater intelligence and power invaded the earth, it was not the high technology of 'superior' humans that saved the world. It was the microbes who decided this intrusion was not to be tolerated, and the powerful alien invaders collapsed and fled before their whirling invisible armies.

If the planet was saved by the grace of God, Wells tells us, then God was manifest in his simplest and most deadly form: the microbe.

3

Early Germ Warfare

Conquistador Y Pestilencia: the European invasion of the Americas • Disease kills 50%–90% • Pizzaro's conquest • Cortez's epidemic kills 10 million • British intentionally spread smallpox • Germ war in Amazon today.

Had the high-technology invaders of H. G. Wells' *War of the Worlds* been more germ-infested than the human race, there would have been no contest at all. There is no defence against superior microbes. This has historical documentation in a one-sided war of extermination that began five centuries ago.

In this confrontation, a race equipped with lethal weapons and even more devastating microbes invaded the lands of a people with no experience of either. This was the European invasion of the Americas. In a very real sense it was a 'war of the worlds', and was even seen as such at the time. The soldiers of the 'Old World' were said to be invading the 'New World' of the Americas.

Geographically protected for millions of years by two great oceans from the continents of Europe and Asia with their belligerent pestilences, the inhabitants of the New World had no natural resistances.

Alfred Crosby Jr., in his book *The Columbian Exchange: the Biological and Cultural Consequences of 1492*, asks a leading question. 'Why were the Europeans able to conquer America so easily? In our formal histories and in our legends, we always emphasize the ferocity and stubbornness of the resistance of the Aztec, Sioux, Apache,

Tupinamba, Araucanian and so on, but the really amazing thing about their resistance is its ineffectiveness.'

There were many reasons, Crosby concludes, but the primary one is betrayed in one of his chapter titles: 'Conquistador Y Pestilencia'. It was a struggle predetermined by the microbes Europeans carried with them. The war for the New World was won by smallpox, influenza, cholera, tuberculosis, plague, malaria, typhus, among other illnesses. Even such minor European ailments as measles and chickenpox proved fatal and killed millions of American Indians.

When Columbus first made contact with the peaceful Carib and Arawak peoples of the West Indies, according to contemporary accounts, the Indians thought the Europeans were divine beings, and 'kissed their hands and feet, marvelling and believing that they came from the sky . . . feeling them to ascertain if they were flesh and bone like themselves.'

If the Indians were seeing the Europeans as a superior race, the true source of that superiority was rooted in a peculiar aspect of their being. Indeed, by virtue of their lack of personal hygiene, their squalid living conditions and their lice and flea-ridden clothing, the Europeans were carriers of superior killer microbes.

In its genesis, North American 'germ warfare' was mostly accidental. Europeans, with superior resistance, simply carried disease to the Americas, then took military advantage when epidemics struck. Disease became the medium of conquest. For instance, Francisco Pizarro reported that a smallpox epidemic had killed approximately 50% of the Inca population shortly before his invasion of South America. Perhaps even more significant was the fact that it decimated the power structure of the empire, wiping out the Inca emperor and virtually all his generals and legitimate successors. This resulted in civil war between factions struggling to fill the power vacuum. Pizarro pitted one side against the other, and managed to crush both. He believed that without the 'blessing' of the epidemic, the Spaniards would never have broken the Inca's power.

In the case of Cortez's conquest of the Aztec empire of Mexico, an epidemic did not precede him. He was, in fact, the means by which it arrived. One of his soldiers, a black man, was a smallpox carrier, and, as the soldier-chronicler Bernal Diaz del Castillo wrote, this man was 'a very black dose for the Indians, for it was because of him that the whole country was stricken'. It was this epidemic that broke the Aztec resistance and eventually killed some ten million Indians in Mexico and Central America.

Throughout the New World, during the first century of contact with the old, the native civilizations were assaulted by a succession of major epidemics which killed 50%–90% of their populations. Several of these epidemics were, in fact, intentionally spread by Europeans. The first historically recorded case of intentional germ warfare in North America was during the eighteenth century, when the British General Amhurst ordered blankets to be taken from a military hospital that was treating smallpox victims. These contaminated blankets were then given out as peace offerings to local Indian tribes. The result, the general later reported, was exactly as he had hoped. An epidemic broke out among the Indians and spread like wildfire among neighbouring tribes, leaving the British troops as the only military force in the region.

In the West Indies, the death rate among the Indians was accelerated by their use as slaves in mines and on plantations. In less than a century, all the tribal peoples of the West Indies – perhaps as many as two million – had been exterminated. It became necessary to bring slaves in from Africa to replace the extinct tribes of the West Indies.

The situation has not radically improved over the last few hundred years. The Amazon river in South America has proved an attractive area for mercantile exploitation during our century. Consequently the surviving Amazonian tribes are being subdued in much the same way as were their North American counterparts in preceding centuries. In their greed for land, white people still find microbes more useful than bullets – although, as several recent massacres have demonstrated, the invaders are not reluctant to use either weapon. Today, as in times past, Europeans are still intentionally introducing killer microbes in order to exterminate tribal peoples and seize their lands.

4

Animal Invaders

Introduction of new animals to Americas • Results: disaster for indigenous species • Invader species: birds in Hawaii; rabbits in Australia; fish in America; mongoose in Caribbean • Chain reactions.

The biological war that the Old World visited on the New World was not, and is not now, limited to diseases which decimated the native peoples. The Europeans brought with them horses, cats, dogs, pigs, goats, rabbits, donkeys, sheep, mongoose and rats. These were all new species in the Americas and, in every case, they displaced or destroyed indigenous New World species.

Carnivorous rats, cats and dogs decimated or – in over a hundred cases – entirely extinguished North and central American birds and mammals. The mongoose was introduced to kill snakes, but it wiped out ground-nesting birds and many mammals and reptiles as well. Pigs, goats and rabbits converted lush tropical islands into deserts. Introduced bird species like starlings and sparrows rapidly drove out many indigenous birds.

In South America, Australia, New Zealand and the Pacific Islands, the story was the same. Europeans introduced predatorial and dangerous species causing cataclysmic chain reactions.

The Hawaiian Islands, for instance, have lost two-thirds of their native fauna because of invading species. Hawaii once boasted the world's most beautiful and varied species of birds – the honeyeaters and honey creepers; as well as many of the world's more remarkable flowering trees. However, this was before Europeans carelessly imported domestic pigeons infected with avian malaria. The pigeon population would have had little effect initially as there were no mosquitoes in Hawaii. In time, however, Europeans brought in mosquitoes as well – including the night mosquito, which serves as the carrier for avian malaria. Rapidly, all of the more than forty species of honeyeaters and honeycreepers that lived below the mountain altitude range in which the night mosquito can exist became extinct. Many also became extinct above that altitude because of deforestation and the introduction of feral cats. Furthermore, as the honeyeaters and honeycreepers were nectar-feeding birds which served, like bees, as pollinators of many large flowering trees, the extinction of the birds will mean the inevitable demise of many species of flowering tree as well.

The devastation of the Australian grassland through the introduction of the European hare is another of the well-documented cases of a disastrously introduced species. The appearance of the fiercely carnivorous European rats in the Pacific and West Indies has also been catastrophic for most indigenous groundnesting birds and small mammals.

Introduced species were not limited to land and air. The accidental introduction of the sea lamprey into America's Great Lakes system by ocean-going ships resulted in the extinction of the lakes' most important commercial fish. The once limitless supply of 'Jumbo Herring' – the blackfin and deepwater cisco – after years of over-fishing, was finally extinguished altogether by the introduced lampreys.

Some introduced species have been strange creatures indeed. In Florida, in the mid-1960s, the 'Walking Catfish' from south-east Asia was imported to fish farms as a novelty. This large, bizarre and very tough fish is capable of climbing out of its pond and moving overland to other pools or streams. By 1969, it had managed to infiltrate almost every pond and waterway in Florida, and push out any and all competitors. When last seen, the catfish was relentlessly marching north and west in its finny conquest of America.

Just how far-reaching the chain reactions caused by introduced species can be is demonstrated in hundreds of recorded incidents. Kenneth Reed Crowell, in *Natural History* magazine, cited the example of the importation into Bermuda, in 1946, of some harmless-looking ornamental nursery plants. Unfortunately these were infested with a scale insect that spread to the indigenous cedar trees. By 1951, 85% of all Bermuda's cedars were dead or dying.

In an attempt to control the scale insect the government introduced ladybird beetles and parasitic hymenoptera, both of which feed on these insects. Unfortunately someone had already imported angus lizards to control two species of previously introduced ants. It turned out that angus lizards, and both species of ants, loved nothing better than to dine on ladybirds and hymenoptera.

Escalating the conflict, the government then decided to attempt to control the lizards by bringing in the kiskadee, a West Indian lizard-eating bird. In 1956, despite protests from conservationists, two hundred kiskadees were released in the hope that the birds would eat the lizards that ate the ladybirds that were supposed to eat the scale insects that killed the cedars. Of course there was a hitch. It turned out that kiskadees found the nestlings of the native Bermuda white-eyed vireo much more to their liking than the scale insects. The plot continues to thicken: the kiskadee has increased to over 100,000 and is decimating the native vireo flocks quite as effectively as the scale insect is ravaging the Bermuda cedar.

The Bermuda problem is typical of those dilemmas faced by environmentalists when dealing with introduced species. It is a continuing biological war creating a never-ending sequence of disasters for indigenous species, a sequence that cannot but ruefully echo the children's song *There Was An Old Lady Who Swallowed A Fly*.

5

The Smallpox War

Edward Jenner's vaccine • Two-hundred-year war • 1966 UN eradication programme • Last gasp 1974 – 25,000 die in India • 1979 – the extinction of smallpox.

In 1796, Edward Jenner developed a safe and effective vaccine against smallpox. So began an epoch in which, for the first time in human history, we were able to do battle with our most formidable enemies on the planet: the microbes. Jenner used the only means possible. He enlisted the help of other microbes to create armies of antibodies which neutralized and destroyed the killer microbes.

Jenner's discovery was a turning-point in human history, a fact that he recognized himself. He understood that his breakthrough had provided humanity not just with a cure for a disease, but with the potential to eradicate it. However, the war to exterminate smallpox was a long one. The disease continued to be one of the major killers of mankind for another century, even though it was much diminished in virulence in those countries where a vaccine was available.

By the beginning of this century it was commonly accepted that Jenner's dream could not be realized. Control rather than eradication was a more realistic objective. In 1966, however, the UN's World Health Assembly voted to create a global Smallpox Eradication Programme: a ten-year plan to eliminate the disease from the planet. The programme was implemented in 1967, when a concerted effort was made to attack smallpox worldwide, and particularly in its endemic reservoir nations in Africa and Asia. There continued to be isolated outbursts of virulence – the most astonishing and vicious being the 1974 epidemic in north-east India, which killed no less than 25,000 people before being fought to a standstill.

By 1975, the disease had been eradicated in India and the world's last case of endemic smallpox was reported in Somalia in 1977. Two years later, the Global Commission certified the eradication of smallpox and its final report for the Certification of Smallpox Eradication was formally accepted at the 33rd World Health Assembly in 1980.

The conquest of smallpox is one of the great success stories of the United Nations and heartening proof of mankind's ability to use biological science for the good of humanity. There are many, many other examples evidenced in the miraculous advances of medicine during the last two centuries.

Regrettably, another faction of the human race, particularly during the last fifty years, has utilized medical knowledge to create new weapons by which the power of the killer microbe is actually increased.

6

Flea, Mosquito and Beetle Bombs

Japanese germ warfare experiments • Three thousand human guinea pigs killed • Sino-Japanese flea war: plague flea-bombs • Russian plague rats • US germ and insect research • Colorado beetle bombs and mosquito missiles • Moroccan Goat Dung Campaign and Anthrax Cattle-cake Option.

In 1940, the Japanese employed the joint weaponry of entomological and germ warfare through the deadly combination of oriental rat flea and bubonic plague bacterium.

Through research conducted by the Japanese army's infamous Unit 731 Biological Warfare Laboratories in Manchuria, some three thousand Russian, Korean, Chinese, American and British prisoners-of-war were used as 'human guinea pigs' and suffered horrible deaths as a result of these experiments. The Japanese military, on the other hand, developed the capability to produce some 500 million plague-infected fleas a year; it also developed an exploding porcelain bomb which could disperse the fleas by air. From October 1940 to November 1941, these insidious flea-bombs were dropped on the cities of Chuhsien, Ningpo, Chanhteh, Kinghwa and at least three other Chinese towns. Outbreaks of plague killed up to a hundred people in each city – areas which had never before known the disease.

However, with horrible irony, the people to suffer most from such weapons systems were the Japanese themselves. Japanese military bungling with their own plague, cholera and dysentery germ warfare weapons resulted in the loss of as many as 10,000 of their own troops!

Unit 731 remained a Japanese military secret until the fall of Japan after Hiroshima and the subsequent American occupation. Rather than bringing the unit's scientists to trial for war crimes, as they should have been compelled to do, US military officials actively protected them and covered up evidence of the existence of Unit 731. Under US protection many of these war criminals went on to lead long and distinguished careers as research and medical scientists – often with the aid of US funds. In exchange, America gained all the poisonous fruits of this secret and horrific research.

The Japanese were not the only nation to explore the potential uses of the flea-plague bomb. The Russians too did extensive research with plague fleas but decided that traditional carriers of the flea were the most effective. According to the Russian ex-army scientist Captain Von Apen, they experimented using 'paratrooper' rats. Von Apen claimed that the Russians had bred highly aggressive strains of grey rat and designed parachute cages to accommodate both the rats and glass phials of plague fleas. Upon impact, the phials would smash, allowing the fleas to leap on the rats who would then be released automatically from their cages to spread disease in the target area. According to Pentagon sources, similar experiments were carried out in the Aral Sea Biotechnical Institute, using squirrels rather than rats. It is also believed that animals accidently escaped from the Aral Institute and this resulted in an outbreak of plague in nearby areas that may have killed as many as two hundred people.

Although it stopped short of using human subjects for experimentation, America had been involved in plague flea research since the beginning of the Second World War. The American military's Entomological and Germ Warfare Department, nicknamed 'The Health Farm', was located at Camp Detrick, Maryland. At one point, feasibility studies there included plans for breeding 50 million plague fleas, ticks with tularemia, and flies with cholera, anthrax and dysentery.

Another line of experimentation was the use of malevolent crop-destroying insects to weaken the enemies' economy and food supply. In retrospect it seems that some of these techniques have actually been utilized. In the autumn of 1944, a plague of hitherto undetected Colorado beetles suddenly appeared in Germany and almost succeeded in wiping out the country's potato crop. Similarly, in 1945, following US air-raids on Japan, massive insect infestations mysteriously appeared and destroyed the rice crops.

America's favourite entomological weapon, however, proved to be the world's leading natural assassin: the anopheles mosquito. By 1954, the Health Farm could produce half a million malarial mosquitoes a month and, by the late fifties, a plant had been developed to produce

130 million a month. The Chemical Corps planned to fire them in cluster bombs from aircraft and in Sergeant Missile warheads.

In 1956, a series of experiments was conducted with uninfected female mosquitoes to see how these specially bred creatures would respond to being fired through the air and exploded in a missile. On their maiden voyage, they were secretly dropped into a residential neighbourhood in Savannah, Georgia. On another test, they were discharged over a Florida bombing range. In both cases, scientists were pleased to see that the mosquitoes adapted well and, within a day, had spread over a two-square-mile area, biting everyone in sight.

After the mosquito and the flea, the bug most predacious on human beings is the common housefly. Its greatest asset to researchers in entomological warfare is that, because of its adaptability, it is potentially the most dangerous of all insects. The housefly can carry some thirty diseases and parasitic worms to man: cholera, typhoid, dysentery, bubonic plague, leprosy, cerebrospinal meningitis, diphtheria, scarlet fever, smallpox and infantile paralysis, to name but a few.

Many experiments exploring the uses of houseflies were carried out by scientists in England, Germany and Japan. One of the most bizarre of these – the Moroccan Goat Dung campaign – was dreamt up in America in 1942 by OSS Commander 'Wild Bill' Donovan's germ warfare expert, Stanley Lovell. During the North African Campaign, when Morocco was occupied by the Germans, Lovell was inspired to enlist the flies of that country as an army of resistance.

Lovell, who Donovan had nicknamed 'Professor Moriarty', had been informed that there were actually more goats than people in Morocco. Lovell therefore reasoned that goat dung was the country's most plentiful resource. With two colleagues, he developed a facsimile dung infected with the bacteria of tularaemia (rabbit fever) and psittacosis (parrot fever), both of which were seriously debilitating illnesses. The artificial dung also contained a chemical that was attractive to flies and would cause them to rouse from hibernation in swarming millions in order to spread the disease to the German troops. Just what awful long-term effects these diseases would have had on the Germans, not to mention the local Moroccans, was fortunately never put to the test. The Germans withdrew from Morocco before the aerial raids could be carried out.

The Moroccan Goat Dung Campaign was similar to a British germ warfare project that was thankfully not made operational either. This was the Anthrax Cattle-cake Campaign which, if it had been utilized, would certainly have resulted in vast areas of Germany being uninhabitable even today. The plan was to drop millions of anthrax-infested, edible cattle-cakes over Germany in order to kill its cattle and spread the highly contagious and horrific disease to the human inhabitants as well. The scientists of the time, however, seem to have had no idea of how deadly and long-lived the anthrax virus was. The island of Gruinard in Scotland, where the project was tested, is to this day totally uninhabitable and contaminated with anthrax. How close the campaign came to being put into operation is evident from the fact that Britain had gone so far as to produce a stockpile of five million deadly cattle-cakes, ready for use.

7

Killer Clams and Others

Shellfish toxins • Cobra venom • Other biological poisons used for espionage and assassination • Invisible weapons in secret war • 'Nondiscernibles' • Markov umbrella-killing and others.

There have been many tales about the killer clam, a huge mollusc which measures up to three feet across the shell and lives for up to a century. According to legend, the clam closes its powerful 'jaws' on the feet of hapless divers who accidentally step into them.

In fact, the great clam, which inhabits the Pacific waters from Taiwan to Australia, closes its jaws so slowly that it presents little real danger to divers. It is the giant clam itself which is in peril. It is being heavily harvested by the Japanese for *sushi* and by the Filipinos to make into bird-baths and curios for the American market. Consequently Indonesia and Australia have found it necessary to protect these creatures by banning their hunting.

In spite of its reputation and proportions, the giant clam is not a killer. However, this is not the case with its smaller relatives. The real killer clam is the edible variety that, during certain times of the year, contains toxins that cause paralytic poisoning in humans. In 1954, Canadian researchers, in co-operation with US scientists at Fort Detrick, managed to refine this toxin in a relatively pure form. It has proved exceptionally effective, working far more quickly, for instance, than cyanide.

Shellfish toxin in Intelligence circles came to replace the cyanide suicide or 'L'pills carried by secret agents. Gary Powers, the American U-2 spy pilot who was shot down over the USSR in May 1960, carried such poison hidden inside a silver dollar. The substance would have worked in ten seconds if he had chosen to take it.

In November 1969, President Nixon announced the United States' intention of destroying all biological weapons. Five years later, however, the Senate Intelligence Committee reported that the CIA had eleven grammes of shellfish toxin in its possession, in addition to numerous other deadly substances, including cobra venom. William Colby, the director of the CIA, explained his organization's use of such toxins in what was called a 'nondiscernible microbioinoculator'. To the unenlightened this looked like a pistol, but Colby explained that the device was electrically powered by a battery in the handle, and, although silent, made an explosion powerful enough to fire a small poisonous dart of shellfish toxin one hundred yards. A gramme of this substance could kill several thousand people. It worked so quickly and effectively that the victim would not even know he had been hit.

Apart from being caught red-handed with deadly contraband, it turned out that the CIA was also guilty of devious methods to avoid detection and censure. While the Watergate scandal was unfolding, the killer substance, along with virtually all other biological weapons developed in the US, was shipped for safekeeping from American arsenals to such places as Porton Down in Britain and its Australian counterpart, the Defence Standards laboratories in Ascot Vale, Victoria, until the political climate improved and the various agencies could reclaim their weapons.

It is commonly accepted that all the major military powers possess biological weapons, and have developed similar 'nondiscernible microbioinoculators'. One such weapon was certainly used in the famous 'umbrella killing' of the Bulgarian journalist Georgi Markov. Markov was murdered in broad daylight outside the headquarters of BBC External Services – Bush House in the Strand – on 7 September 1978. Before his death Markov, who had some knowledge of the technique involved, was just able to communicate the fact that he was sure the umbrella was an assassin's weapon.

When Markov's body was examined, a tiny pellet the size of a pinhead was discovered embedded in his leg. There were four tiny holes in the pellet and these had released the deadly poison, which was at first thought to be shellfish toxin. However, it turned out that the KGB, Markov's assassins, were utilizing killer plants rather than killer clams; the poison proved to be ricin, a toxin derived from the seeds of the castor oil plant.

A similar, but unsuccessful, assassination attempt occurred in the Paris Metro when another Bulgarian exile, Vladimir Kostov, felt a sudden sharp pain in his neck. Fortunately for Kostov, the poison was either old or in diluted quantities, for it did not kill him. A pellet similar to that found in Markov's leg was extracted from Kastov's neck.

The number of people who have met their end through an indirect confrontation with the killer clam and its allies is anybody's guess. In the secret world of espionage, most of the killing that takes place is attributed to natural causes.

8

DNA and the Genetic Bomb

US germ war experiments • Near disasters •
Biological weapons ban of 1972 • bioengineering
• Designer diseases without cures • 'Supergerm'
and ethnic plague bombs.

Since the Second World War, most of the world powers have denied that any germ warfare experiments have been conducted in their countries. It would be naive and even dangerous to accept this affirmation in the light of available evidence documenting the proliferation of this insidious form of warfare.

In 1977, a senate committee forced the revelation that the US Army, between 1949 and 1969, had conducted no fewer than 239 'open air' germ warfare tests. During one week in 1950, the Army subjected the city of San Francisco to six biological warfare assaults. Military officials claimed that the bacteria were harmless, but scientists pointed out that one of the micro-organisms was *serratia marcescens*, which can be fatal to elderly persons and newborn infants, as well as to those with heart and lung diseases. At least one person died as a result of this episode and it is claimed that at least eighteen people were treated in San Francisco for an outbreak of a rare flu-like disease brought on by the bacilli.

Undeterred, in 1966 the military conducted a similar experiment in the New York subway system during rush hour. Again the bacilli were declared benign, but proved to be otherwise.

In a third example, in 1952, a combined British-American-Canadian naval exercise off the coast of Scotland resulted in an airborne 'biological bomb' of plague bacilli being accidentally sprayed on a civilian fishing trawler, the 400-ton *Carella*. Only a failure in the effectiveness of the airborne system's ability to carry the disease saved the lives of the fishermen.

The sobering truth is that, with current advances in biological weapons, we now have more to fear than just the threat of a global nuclear war.

Following President Nixon's declaration that the United States was destroying its stockpile of biological weapons, the Biological Weapons Convention banned their development, production and use in 1972. Remarkably, this was the first treaty to actually result in the riddance of a life-destroying arsenal. A decade later, however, America was spending 30 million dollars a year in 'medical defences against biological weapons'. (This is a very old ploy: even the blatantly offensive and infamous Japanese Unit 731 went by the name of the 'Epidemic Prevention Unit'.) Anyone with any expertise in this field knows that there is no such thing as a strictly defensive biological weapon.

So what is responsible for the virtual abandonment of this important treaty? The answer is bioengineering.

The discovery, three decades ago, of the DNA structure by James Watson and Francis Crick meant the advent of genetic engineering; biological science's equivalent to splitting the atom.

Advances in the technology of bioengineering during the last decade, based on the Watson-Crick discovery, have opened the door to a new science where almost anything seems possible. By tinkering with the genetic structure of life itself, it is possible to alter life-forms, as we have already seen. Bioengineering is perhaps our best option for finding elusive cures for complex diseases like cancer. It is a discipline that will revolutionize the medical profession, the chemical industry and agriculture. Unfortunately, it has another, darker side: the Frankenstein factor. It can be used as a tool to create monstrous weapons in the fields of biological warfare.

Bioengineering also opens up the possibility of creating 'designer diseases' – through the genetic alteration of bacteria – for which there is no cure.

A more sophisticated and likely development in biological warfare, however, would be the use of recombinant DNA techniques which could reinvent diseases like diphtheria, plague, yellow fever and smallpox in new and deadlier forms. Scientists could conceivably produce a protective vaccine to safeguard themselves and their allies against their own abominable inventions. Another deadly alternative is the so-called 'ethnic bomb', a Nazi dream weapon. It is possible that scientists could create viruses to which certain racial groups – such as blacks or orientals – are vulnerable but which leave others untouched.

The pursuit of the Super Germ has gone on for some time. Scientists at the US Army's biological warfare headquarters at Fort Dietrich were working continuously on genetic research from 1942 to 1969, when President Nixon stopped its programme of research and weapon stockpiling.

In 1979, when the laboratories at Fort Dietrich reopened their biotechnology programme, the head of biological research, Dr William Beisel, announced: 'All the work we are doing is defensive'.

9

Chemical Warfare

Consequences of chemical pollution

'All that is necessary for the triumph of
evil is that good men do nothing.'
Edmund Burke

1

Sanctions: Ecological Assassination

Murder of Italian executive for Seveso dioxin disaster • Contamination of 11,000 people • Killing of 40,000 animals • Court actions.

On a February morning in 1989, Enrico Paoletti, a thirty-nine-year-old corporate executive from Monza, near Milan, rose early for work as usual. As he walked briskly from his home to his car, he was suddenly confronted by three men and a woman carrying automatic weapons. The attackers opened fire, killing Paoletti instantly.

Within hours of the assassination, the left-wing terrorist group Prima Linea, or Front Line, claimed responsibility. A few days later, in leaflets circulated in a Turin outdoor market, Front Line disclosed the reason for this assassination by one of its guerilla commando units. Paoletti was the industrial director of the Swiss-owned Icmesa Chemical Company, a subsidiary of the Hoffman-La Roche group. He was also the director of Icmesa's northern Italian plant in Seveso where, in 1976, an explosion had resulted in the disastrous release of a dioxin cloud that killed virtually all the town's domestic animals and caused severe skin disease in many adults and children.

Subsequent investigation showed that the 1976 incident at Seveso was clearly the result of negligence on the part of the company. Paoletti and a few other executives were placed under arrest after the disaster, reputedly Italy's worst ecological mishap. After six months in prison Paoletti was released on bail pending a trial which never took place. The Front Line guerillas, deciding that justice would never ultimately be served, took it into their own hands to deliver what they believed was just punishment for the businessman's crime against the people of Seveso.

The explosion at Seveso's Icmesa chemical plant on 10 July 1976 set off a deadly chain of events that resulted in immediate measurable effects on plants and small animals. In spite of this, the regional government took ten days to declare the area polluted by dioxin and to evacuate the eleven thousand citizens. By this time forty thousand farm and domestic animals were dead from the effects of the poison cloud. More than three hundred houses were demolished and buried with soil trucked in from uncontaminated areas. The heart of Seveso was destroyed, perhaps forever.

Over four hundred children suffering from chloracne, a disfiguring dermatological disease manifesting itself in recurring boils and skin eruptions over face and body, were hospitalized. The same number of expectant mothers were aborted and the expectation is that the rate of cancer in the town will increase dramatically.

What is more shocking than the effects of the accident is the inadequate and delayed response of the authorities. No one in Seveso seemed to realize the magnitude of the disaster. Dioxin is seven thousand times more deadly than cyanide and not even the factory workers were fully aware of the toxicity of the chemicals they handled. Furthermore, the industrialists refused to accept responsibility. A leading executive attempted to minimize the catastrophe in a fatuous statement that only added to the enragement of the victims. 'Capitalism means progress,' he claimed, 'and progress can sometimes lead to inconvenience.'

This unbelievable insensitivity to the realities of Seveso fuelled an already volatile political situation in Italy and led to the first ecologically motivated assassination. Typically, the company lawyers were able to delay final judgment for ten years, and although five executives were initially given two-and-a-half to five-year prison sentences for wilful negligence, by the time the appeals were heard, all but two were acquitted. Neither of these spent more than one year in jail.

2

A Chronicle of Disasters

Halifax, Canada, 1917–2,000 dead • Oppan, Germany, 1921–560 • Hawks Nest West Virginia, 1931–35–2,000 • Texas City, US, 1947–600. Ludwigshafen, Germany, 1948–200 • San Carlos, Spain, 1978–200 • Salang Tunnel, Afghanistan, 1982–2,000 • Cubatao, Brazil, 1984–600 • Ixhautepec, Mexico, 1984–1,500 • Bophal, India, 1984–3,000–10,000 • Mindora, Philippines, 1987–2,000 • North Sea, UK, 1968–88–500.

The Seveso disaster was just one of thousands of large-scale industrial accidents that have occurred this century. With hindsight – or even a minimum of foresight – nearly all of them were preventable.

Below are listed some of the worst industrial disasters of the twentieth century. As will be seen, despite claims by all major corporations that safety standards in industry are continually improving, industrial accidents seem to be taking more and more lives as the century moves on. 1984 proved to be the most ill-fated year in history in this respect, with major disasters in Mexico, Brazil and India.

6 December 1917 – Halifax, Canada
The Belgian relief ship, *Imo*, grossly overburdened with unstable and outdated explosives, accidently rammed the French munitions vessel, *Mount Blanc* in the harbour of Halifax, Nova Scotia. The resulting explosion destroyed half the city.

Between 2,000 and 3,000 were killed and 8,000 were seriously injured in the disaster. Railway carriages were hurled hundreds of yards off their tracks. The suburb of Richmond was entirely obliterated. In one freak incident a Halifax man was hurled through the air and landed, uninjured, in a tree over a mile away.

21 September 1921 – Oppan, Germany
The biggest chemical explosion in German history occurred in a warehouse in Oppan just south of Frankfurt. Foolishly advised workers used dynamite to blast loose 4,000 tons of ammonium nitrate fertilizer. In the blast, the ammonium nitrate itself converted to an explosive, and the entire load became a massive bomb. The death toll was over 560, with at least 3,000 serious injuries. A large section of Oppan was subsequently demolished.

1931–1935 – Hawk's Nest, West Virginia, USA
Under direction of the Union Carbide Company workers constructed a hydroelectric tunnel at Hawk's Nest. Ventilation safety was largely ignored throughout the tunnelling operation resulting in the worst casualty rate through death by silicosis in history. It is believed that between 2,000 and 3,000 workers died from silicosis acquired while building this tunnel.

16 April 1947 – Texas City, USA
The freighter *Grand Camp*, anchored in Galveston Bay, caught fire. Unknown to port authorities it contained 1,500 tons of ammonium nitrate fertilizer, the same deadly substance that destroyed Oppan. At 8.00 a.m. the smouldering ship exploded with such force that its impact was registered a hundred miles away. But the disaster did not stop there. Fire engulfed the Monsanto styrene synthetic rubber plant in the harbour, which in turn exploded, spreading fire throughout the town. This fire burned on until the following day, when a second freighter, the *High Flyer*, also loaded with nitrates, caused a third disastrous explosion. The death toll was nearly 600, with approximately 2,000 serious burn victims.

28 July 1948 – Ludwigshafen, Germany
At the factory site of the I.G. Farber chemical plant, a railway car containing the highly volatile dimethylethen was carelessly ignited by workers who were not told of its contents. The resulting explosion destroyed the factory and all its workers; it also caused a firestorm in the town, killing 200 and injuring 4,000.

11 July 1978 – San Carlos de la Rapita, Spain
An overloaded tanker, carrying combustible propylene gas and travelling too fast, slid out of control on a corner and struck a wall. The result was a raging sheet of flame that engulfed a roadside campsite

acccommodating nearly 800 tourists. Over 200 campers were burned to death in the inferno.

3 November 1982 – Salang Tunnel, Afghanistan

A petrol tank truck exploded inside the two-mile long Salang Tunnel. The Soviet and Afghan military mistook the collision for a guerilla army attack and sealed off both ends of the tunnel. The result was that fire and carbon monoxide poisoning killed approximately 2,000 people trapped in the tunnel – although some put the figure as high as 2,700.

25 February 1984 – Cubato, Brazil

Seepage from a leaking petrol pipeline in the south-east Brazilian town of Cubato was allowed to accumulate, despite complaints of fumes. When the leak was finally ignited, the explosion resulted in a giant fireball that razed the town to the ground. At least 600 died and 3,000 sustained injuries.

19 November 1984 – Ixhautepec, Mexico

At 5.43 a.m. the Mexico City shanty town suburb of St Juan Ixhautepec was rocked by a massive explosion that enveloped it in flames. This was the initial blast of an hour-long series of disturbances registering .5 on the Richter earthquake scale. Fireballs measuring 300 metres in diameter roared through the streets for nearly two hours, destroying entire blocks of houses and incinerating every form of life in their path.

The reason for the explosion was careless storage of liquified petroleum gas, mostly in the form of propane, at the state-owned Pemex Oil Company's gas complex in Ixhautepec. Pemex operations in Mexico had previously been described by inspectors as 'a catastrophe waiting to happen'. In the six months prior to the disaster, there had already been four separate major incidents, resulting in 89 deaths and hundreds of injuries. It was days before the last fires at Ixhautepec were extinguished and the casualties estimated. Well over 1,500 died in the disaster, and at least 7,000 were seriously injured.

2 December 1984 – Bophal, India

When the industrial bomb exploded at Bophal, some five hundred miles south of New Delhi, there was no thunderclap and wall of flame. It happened silently, invisibly in the dead of night. For men, women and children in this town of a quarter of a million people, the end of the world came without a bang or even a whimper. Through a series of errors and technical faults the American-owned Union Carbide pesticide plant in Bophal allowed 30 tons of the deadly toxic gas methyl isocyanate (MIC) to escape from storage tanks and drift through the night air into the sleeping town.

Many died without knowing it, others – as far as twenty-five square miles away from the plant – woke with burning pain in the eyes and lungs. Many found they were totally and permanently blind, others were gagging on the air but did not realize that their lungs were critically damaged. It will never now be possible to determine exactly how many died at Bophal, but the figure is certainly not less than 3,000 and may well be nearer 10,000. Over 200,000 people were exposed to this gas attack, nearly half of whom were evacuated. At least 20,000 were permanently disabled by the disaster.

12 December 1987 – Minedoro Island, Philippines
The greatest ferry and peacetime ship disaster in history was caused by the oil tanker *Vector* when it struck a heavily loaded ferry off Mindoro Island in the Philippines. The Manila ferry *Dona Paz*, owned by Sulpicio Lines, sank rapidly with the loss of between 2,000 and 3,000 passengers. Coastguard reports clearly blamed the disaster on the oil tanker's actions.

2 July 1988 – North Sea, UK
The world's worst oil rig disaster occurred on Occidental Petroleum's *Piper Alpha* platform in the North Sea. The explosion and fire claimed the lives of 166 oil workers.

In 1980, the Norwegian platform station, *Alexander Kielland,* capsized in the Ekofisk field in the North Sea, killing 123. In November 1986, 45 oil workers were killed just off the Shetlands in a helicopter crash.

While the North Sea has the worst record of any single region for oil field disasters, with a death toll of more than 500, hundreds of oil workers have died in other regions during the last decade. An American oil-drilling ship sank in the South China Sea in October 1981, killing 81. The following year, in September 1982, another American-owned oil rig, *Ocean Ranger,* keeled over in the North Atlantic and 84 men perished.

3

Toxic Time Bombs

Toxic waste in US • 80 million tons per year; 90% illegal dumping • 50,000 sites • Environmental Protection Agency warnings • Love Canal and other disasters.

Despite their dramatic impact, large-scale industrial disasters are relatively rare events in the overall theatre of chemical war. It is the secondary disasters, those occurring without immediate effect and without the photographs and body counts, that are the most frightening and difficult to deal with in terms of regulation and control.

The American Environmental Protection Agency (EPA), which came into being in 1970, has labelled toxic waste as 'the most serious environmental problem in the US today'. America alone produces 80 million tons of hazardous waste each year – the equivalent of more than 700 pounds per person. Further, the EPA points out that 90% of all hazardous waste is dumped illegally or in a manner likely to threaten life.

The EPA recognizes some 50,000 hazardous waste sites in the United States, of which at least 2,000 may pose 'significant risks to human health or the environment'. Further, they note, only 200 of these sites were even licensed for disposal of toxic waste.

An EPA official stated the case bluntly: 'Every barrel stuck in the ground is a ticking time bomb.' The cost of the clean-up after years of industrial dumping is almost unimaginable. One study estimates that the cost of cleaning up just one thousand of the most dangerous would be 44 billion dollars.

The most infamous and emblematic toxic waste case is that of Love Canal near Niagara Falls, New York. Love Canal and its proximity to the natural wonder of the great falls has become a metaphor for the dark side of the American dream. Love Canal was a failed canal project built by William Love in the 1890s. The abandoned, canal trench was some three thousand feet long, sixty feet wide and ten feet deep and for a time became a popular swimming spot. However, during the 1940s the Olin Corporation and Hooker Chemicals, the owners of the site, began using the Love Canal to dispose of toxic chemicals. From 1942 to 1952, Hooker dumped some 21,000 tons of toxic chemicals into the canal.

In 1953, in what appeared to be an act of philanthropy, Hooker disposed of the dump site by deeding the land for one dollar to the local Board of Education so that a school could be built there. The company failed to warn the board about the nature of the deadly chemicals in the dump. They did, however, have the forethought to protect themselves by writing a disclaimer into the deed, which transferred liability to the school board for any claims that might arise in connection with the buried toxic waste.

In this way, Hooker was simply unloading responsibility for a toxic waste site on to an unknowing community. Despite the fact that, as early as 1958, there is documentary evidence to show that Hooker was aware that children from the school were being 'burned by material at the old Love Canal property', the company still did not reveal the truth about the site to the school board or those with homes in the area.

It was not until 1976 that the community awoke to the horror of the toxic poisoning. After several years of unusually heavy rains, the water table in the area rose and basements and house foundations began to stink of chemicals. Entire gardens withered, pets died and children developed severe chemical burns on their hands and feet. Surveys soon indicated that air, water and soil pollution in the area was several hundred times above the safe level. Also the population was showing extremely high levels of birth defects, miscarriages, cancers, blood diseases, hyperactivity, asthma, nervous breakdowns and suicides.

Despite overwhelming evidence, spokesmen for the chemical industry attempted to discredit the claims of residents at Love Canal, denying all responsibility and even attempting to blame the victims themselves. In a television interview, the vice-president of the Chemical Manufacturers' Association took the official industry line and implied that the residents of Love Canal were simply a community of hypochondriacs. At times, even many of the government bodies that were supposed to protect citizens were reluctant to deal with such a troublesome issue. It was only because a strong community-based residents' organization forced the issue that anything was achieved at Love Canal.

In 1979, when only two out of seventeen pregnant women in the community had managed to give birth to healthy children, New York authorities termed Love Canal an area of 'grave and imminent peril'. The school was closed, more than two hundred houses were demolished and over a thousand families were evacuated at a cost in excess of $100 million.

As nightmarish as the situation at Love Canal was, it is one which will recur again and again throughout the industrial world. The EPA claimed 'We know of 4,000 to 5,000 potential Love Canals right now'. Hugh Kaufman, head of hazardous waste assessment at the EPA, said, 'the only unusual thing about Love Canal is that it was discovered.'

4

Poison Food

Rachel Carson's *Silent Spring* and DDT Poisoning • Pesticides: 3 kilos per person annually • Oxfam: 1 million poisonings, 5,000 deaths annually • 1973 Michigan contamination • Iraqi poisoned grain, 1971–6,000 dead • European wine scandal • Spanish olive oil poisonings, 1981–350 dead.

In 1963, *Punch* magazine upheld its reputation for biting satire by publishing a black joke in the form of a cartoon portraying a man standing next to a rigidly dead dog. The caption read: 'This is the dog that bit the cat that killed the rat that ate the malt that came from the grain that Jack sprayed.'

As popular humour often does, it told a great deal about a very recent awareness on the part of the public of the effects of pesticide poisoning. To a large degree this understanding of the chain reaction of events triggered by chemical treatment of crops was due to Rachel Carson's landmark book, *Silent Spring*, published in 1962, which revealed some terrible truths about the poisonous and carcinogenic nature of DDT and other chemicals used by the agricultural industry.

Although viciously attacked by the industry and accused of being part of a communist conspiracy, Carson's book *finally led* to the banning of DDT in the USA and many other countries. However, as critics continue to point out, the issue did not end there: America continues to export the substance to Third World countries which, in turn, use it on produce that is shipped back to the United States – and thus DDT remains a basic part of the average American diet.

The sad truth is that all the chemicals named in *Silent Spring* continue to be manufactured, and many more even more deadly ones have been added to the range. The chemical industry produces the equivalent of three kilogrammes of pesticide for every man, woman and child on the planet each year. In America alone it is an $8-billion industry.

Each year OXFAM reports nearly a million cases of farm workers suffering from pesticide poisoning, and at least five thousand deaths a year are a direct result of this. In fact, most environmentalists believe the casualty rate is more than twice the recorded figures, as only a percentage of poisonings are ever reported in the many poor areas of the Third World where pesticides are extensively used. Furthermore, these figures do not reflect the damage that pesticides and other chemicals are known to cause through secondary diseases like cancer. Even though regulations governing the use of chemicals in food production are more stringent in the United States than almost anywhere else, it is obvious that these chemicals are indirectly linked to what has been labelled a cancer epidemic, which strikes one in four Americans. The casual use of highly toxic pesticides has also led to a large number of mass poisonings. A classic example occurred in 1973 in Michigan when a truck driver collected a ton of the innocuous feed additive called Nutrimaster from the Michigan Chemical Corporation for delivery to the largest agricultural feed plant in the state. The ton of white powder was mixed into the feed products and shipped out to farmers all over Michigan.

Tragically, an error had been made. The white powder was not Nutrimaster, but Firemaster, a highly toxic flame retardant. This toxic mix poisoned tens of thousands of animals, and the deadly chemicals filtered through to virtually every food outlet in the state. The result was that some nine million people ingested a severe overdose of the chemical PBB – which is known to provoke cancer and cause genetic damage in humans. The ultimate toll of this mass poisoning will take generations to estimate.

The most deadly of all chemical poisonings of food supplies, however, occurred in 1971. During that year, in order to relieve a famine, the Iraqi government purchased nearly 100,000 tons of seed grain. Unfortunately, the grain had been treated with methylmercury dicyanidamite – a mercury-based chemical banned in Europe and North America.

This grain was distributed widely to Iraqi farmers for sowing the following year. Warnings of its poisonous chemical content were written in English and Spanish on the sacks, but as few Iraqis understood either language and translations were not provided, the warnings went unheeded. The situation was further complicated by corrupt Iraqi officials who stole thousands of tons of the grain. Removing it from its sacks, they sold it for food on the black market.

The result was mass mercury poisoning for those who either ate the grain or fed it to their animals. It is difficult to estimate the scale of the damage because government reports were both incomplete and intentionally falsified to minimize the scandal. Unofficial estimates suggest that about six thousand people died and perhaps one hundred thousand were blinded, paralyzed or made severely ill through mercury contamination.

Accidental poisonings are one thing; the deliberate contamination of foods quite another. This criminal activity has become increasingly common in recent years. The standard bar joke of 'pick your poison' turned sour during 1986, when it was learned that European wine-makers were routinely 'fortifying' poor-quality wines with chemical spiking ingredients. In Italy that year a particularly badly spiked batch of wine actually killed thirty people. An even more shocking revelation at this time, however, was that virtually all the major 'respectable' wine producers in Austria, and a few in Germany, were spiking their wines with the chemical ingredient of antifreeze. This unscrupulous doctoring of wines was engineered by the chemist Otto Nadrasky Sr. In a scandal which resulted in the seizure of virtually all Austrian wine stocks, and the loss of the annual $30-million Austrian wine export business, Nadrasky and thirty others were finally brought to trial for their outrageous fraud.

It emerged that the practice had been an industrial secret since 1977, when Nadrasky discovered that diethylene glycol (antifreeze) gave cheap wines more body and sweetness. More importantly, it could not be detected by the usual chemical tests conducted on wines. The fact that habitual consumers were likely to contract severe kidney damage did not seem to deter wine producers bent on maximizing profits.

Such widespread tampering could not have been sustained without political support, and as the scandal unravelled, it became apparent that there had been considerable corruption among government officials. In the end it was only the institution of a new testing system for wines that led to the discovery of the spiking ingredient.

The worst case of food-tampering in history occurred in Spain, in 1981, resulting in the deaths of over three hundred and fifty people and the severe poisoning of thousands more. In May of that year hundreds of Spaniards in villages just outside Madrid were being admitted to hospital with fever, backache and lungs full of fluid. Young men were suddenly turned into old men suffering from pneumonia. Entire families were wasting away and being afflicted with paralysis. Health workers could not understand what was causing this withering plague. It took over six weeks to discover that the source of the deadly epidemic was olive oil. This oil resulting from overproduction of EEC quota figures, had been converted into industrial oil by mixing it with rape seed oil which contained a poisonous analyne dye to make it inedible. A large quantity of it was then acquired cheaply by black marketeers for sale on the domestic market. This necessitated removing the poison dye, during which process an alteration in the molecular structure of the oil produced an insidious poison that was not apparent for weeks after consumption. This poison attacked the human central nervous system with devastating effect; it led to paralysis of nerves and muscles that proved impossible either to identify or treat.

Some twenty-six olive oil black-marketeers were arrested following the disaster, but it was not possible to trace the source of the oil because of the complexity of the operation. The mass killers went unpunished, and the inquiry into the matter was closed in a mood of outrage and frustration in 1985.

5

Poison Air

London's 1952 killer fog: 4,000 dead • US Clean Air
Act 1970 – loopholes • 53,000 US air pollution
related deaths • Mexico's killer smog • Politics of
pollution.

By the 1950s, London was being subjected to 'pea-soup' smog attacks due to uncontrolled industrial and domestic air pollution. Common sense had long dictated that measures must be taken to alleviate the pollution problem, but it took an outright disaster to motivate the government to act for the common good rather than for the profit of powerful industrial lobbyists.

Such a disaster was long overdue when, in 1952, a pea-soup fog of unprecedented density enveloped London for three weeks without lifting. The smog was so thick that it was virtually impossible to tell day from night and the city was almost brought to a standstill. The high concentration of sulphur-dioxide in the smog, resulting from industrial pollution, brought on fatal coughing fits among the very young and the very old. When the air finally cleared, it was discovered that during the three-week period when the 'killer fog' had held London in its grip, four thousand people had died of lung-related illness brought on by the fog.

Strong public reaction to the disaster led to the formation of the emergency Atmosphere Pollution Committee. This agency had wide-ranging powers. For the first time, reasonable anti-pollution controls were put into effect and London was made a smoke-free zone. It was the beginning of an industrial clean-up long overdue in urban industrialized Britain.

American cities like Los Angeles and Philadelphia were, by the 1960s, suffering smog problems as severe as London's, and national environmental efforts resulted in the drafting of America's Clean Air Act of 1970, perhaps the most sweeping set of anti-pollution laws ever drawn up. Although the Act fails to address the issues of acid rain and airborne toxic chemicals, it is essentially aimed at the common good.

Despite the fact that some environmental surveys conclude that 53,000 Americans a year die prematurely from lung ailments brought on by air pollution, industrial lobbyists have done everything in their power to weaken the Act. The Reagan administration in particular went to considerable lengths to minimize its effectiveness.

If air pollution is considered a major problem in the cities of North America and Europe, the newly industrialized large cities of Latin America, Asia and Africa can only be described as operational disaster areas. It is obvious, for instance, that the world's largest untamed metropolitan sprawl which goes by the name of Mexico City would, if properly monitored, reveal that more people die there every year from killer toxic fumes than died in London during the fatal 'killer fog'. Entire flocks of birds are commonly found dead in the streets simply from inhaling the toxic mix of chemicals that passes for air in parts of Mexico City.

Depressingly, for places like Mexico City, there seems no solution in sight. In the face of political corruption and the uncontrolled urban growth that is likely to expand to an estimated twenty million by the end of the century, the best advice the City Council has offered those suffering lung ailments is: 'Try to breathe at all times through your nose'. Not very sound medical advice, perhaps, but as another way of saying 'Keep your mouth shut', it may serve its purpose.

6

Gas Warfare

Ypres, 1915—first gas attack • Dr Haber's 'higher form of killing' • WW I 90,000 dead; 1,300,000 casualties • Nerve gas • Agent Orange • Iran-Iraq—5,000 dead • Terrorist's dream: VX nerve gas—$5 per million man-lethal dosages.

On 22 April 1915, in a field near Ypres in France, chemical warfare made its macabre debut. German soldiers opened the valves on six thousand cylinders of liquid chlorine along a four-mile front, and a gentle wind wafted the gas across to the French troops. Five thousand French and Algerian soldiers died and ten thousand were disabled in the attack. Two days later, the scenario was repeated and five thousand Canadian troops were killed in a second gas attack.

Strangely enough, Fritz Haber, the German chemist who is the acknowledged 'father of chemical warfare' and masterminded these unprecedented attacks, was not tried for war crimes at the end of the Great War. Instead, in 1919, he received the Nobel Prize for Chemistry and in his acceptance speech, he had the audacity to flaunt his atrocity. 'In no future war,' boasted Haber, 'will the military be able to ignore poison gas. It is a higher form of killing.'

More shocking than the aberrant behaviour of one chemist is the general toleration of his views by governments. Gas warfare in the First World War resulted in the death of 90,000 troops and the crippling injury of 1,300,000 others. Many of these suffered for what was left of their lives from the horrific effects of chlorine and the even more deadly phosgene – the notorious 'mustard gas'.

Nor did the research into poison gas end with the discoveries of Haber. In 1937, while researching insecticides, Dr Gerhard Schrader and others developed a gas that proved to be even more deadly than mustard gas. This was a gas called 'Tabun', which came to be known as 'nerve gas' because it attacked the human nervous system. Shortly afterwards, the research team developed a second nerve gas, 'Sarin', which was ten times again more dangerous than 'Tabun'.

During the last war the Nazis could certainly claim a superiority when it came to chemical weapons for, in addition to the two nerve gases, they had developed a superior mustard gas and a new incendiary gas – all of which could be delivered in their notorious V-rockets. However, it seems that Hitler, who had experienced gas attacks in the First World War, had an inherent dislike of the weapons. Moreover, he had found them unnecessary for success at the beginning of the war, and in the late stages, he lacked the superior air support necessary for their successful deployment.

Chemical warfare played a larger part in the American-Vietnam conflict than it did in the Second World War. American bombers and soldiers used CS gas bombs against the Viet Cong, especially in their underground tunnels and bunkers. Even more widespread was the use of defoliation chemicals – like the notorious Agent Orange – to destroy crops and wipe out entire jungle areas that offered cover to the enemy. By the end of the war, Americans had sprayed over 240 pounds of dioxin over South Vietnam. This may not sound like a large amount, but dioxin is so toxic that two ounces of the substance in New York City's water supply would be sufficient to kill every inhabitant. Both the country and the people of Vietnam will bear the scars of that war for a century or more. Ironically, many Americans are also inheriting a small part of this legacy of cancer and genetic mutation simply because they handled the deadly chemicals.

The American and Soviet armed forces have built up massive chemical warfare stockpiles of increasingly dangerous substances, as have many other nations. It is obvious that some of these weapons were deployed by the Soviets in the recent Afghan war, and that some other Western nation supplied Iraq with chemicals with which to fight Iran. President Saddam Hussein has used chemical weapons on at least sixty occasions since April 1987, but in most cases these have been employed within Iraq against the Kurdish ethnic minority in what appears to be a policy of racial genocide. March 1988 saw the worst single atrocity in this cowardly and reprehensible campaign against defenceless civilians. Iraqi forces attacked the Kurdish town of Halabja with poison gas. Within minutes of the air strike the entire population of some five thousand Kurdish men, women and children was dead.

Apart from major powers using and sharing their arsenals of genocidal chemical weapons, the use of these weapons by terrorists has raised its head in recent years. Joseph Douglas, author of *C.B.W.: The Poor Man's Atomic Bomb*, claims that 'most of the stuff you need to build chemical and biological weapons is available through the mail. Chemical and biological "Saturday Night Specials" increase the terrorist capability immensely. There is really nothing to stop anybody from creating a toxin. It's literally something you can do in a garage.'

In a post-war update on Nazi nerve gases, the most lethal of these is now a substance known as VX gas. Given the formula, VX gas can be made by a second-year chemistry student. A quart of VX gas, according to the journal *Science*, costs about five dollars to manufacture, and contains several million man-lethal doses. In 1971, US and British scientists put the essential components of VX gas into the public record at the Geneva Disarmament Conference. Consequently the frightening possibility exists that this terrible weapon is freely available to any terrorist with sufficient initiative to walk into any major library in the world and spend half an hour researching the subject.

7

Poison Water

Water pollution: 25 million deaths annually • 50% of world population lacks clean water • Example of Bogotá • UN Decade of Water 1980–90 • Taps and toilets for world • Toxics in aquifers • Sea pollution: oil spill disasters, toxic waste, sewage • Minamata Disaster • UNEP efforts to save the seas.

Polluted water is the single greatest cause of human illness and death through disease. Half the world's worst diseases are spread through polluted water. Half the people in the developing world don't have access to clean water. Three-quarters do not have acceptable sanitation. Health workers in developing countries are in agreement that the number of taps and toilets in a nation are a far greater indication of its people's health than its number of doctors and hospitals. The fact is, polluted water leads, in one way or another, to the death of some 25 million people annually.

Examples of criminal disregard for safe water supplies can be found everywhere, but many of the most outrageous cases have occurred in the Third World. The Bogota River in Columbia is a typical example. This river is almost the sole source of water for approximately five million people. It is also used by the country's largest industrial complex as an open sewer, with the result that all towns downriver from it are forced to use contaminated water for their domestic and agricultural needs. Consequently infant mortality, diarrhoea and gastroenteritis rates on the lower river are the highest in Columbia. In one town directly below the industrial complex the infant mortality rate – already the highest in the country – has doubled in the last ten years. This increase exactly matches the expansion rate of the industrial complex over the same period.

Through the 1970s the issue of taps and toilets in the Third World became a major area of debate and concern in the United Nations. Admirably, 1981–1990 was declared the 'United Nations Water and Sanitation Decade'. The World Health Organization set itself the goal of clean water and access to basic sanitation for all the people of the world by the end of the decade.

Although this scheme would involve supplying well over three billion people with taps and toilets, the costs were not impossible. UN estimates suggested that an expenditure worldwide of between $30–60 billion for each year of the decade would achieve the WHO's goal. This amount would be the equivalent of between 4% and 8% of the world's annual military budget. Furthermore, the WHO pointed out, for most nations such an investment would almost immediately produce equivalent savings through preventative medicine. Far less would have to be spent on hospitals and medical supplies, disastrous epidemics would be less likely and the workforce would be healthier and so more capable of producing wealth for the nation.

Unfortunately the UN's Water and Sanitation Decade has proved more of an inspired dream than an achieved programme. The WHO revised its goal of taps and toilets by the end of the eighties to a still very respectable one billion people in the Third World, but even this has proved beyond its capability. Self-interest and political expediency among international politicians has resulted in the project being massively underfunded.

One should not, however, be dismissive of the ideals and achievements of the UN. Ten billion dollars a year will in fact have been spent worldwide on water and sanitation projects during the decade. Moreover the WHO programme has been extraordinarily successful in many Third World countries in educating both leaders and general populations on the relationship between water and sanitation and a nation's health. There is hope that this knowledge will result in considerable change in the long term.

Although the worst examples of water pollution are certainly to be found in the Third World, no one should delude himself that the problem only exists in developing countries. In 1980, a two-year research programme conducted on Long Island by the New York Public Interest Research Group (PIRG) was released. The report, which was called 'Toxic's On Tap', listed some 66 companies which were dumping nearly ten million gallons of polluted waste water into Long Island's eleven sewage systems each day. None of these systems was capable of treating toxic waste. Hooker Chemicals, of Love Canal fame, was named as one of the area's major polluters, especially with carcinogenic chemicals. These companies had, according to the report, permanently contaminated the underground aquifers which were the only source of water for Long Island's three million residents.

One of the authors of the report emphasized the gravity of the situation by pointing out that even if all pollution was stopped tomorrow, it would take three thousand years for the pollutants to bleed out of the water table. And as New York has some of the strongest legislation in America to control water pollution, it is frightening to imagine what the water pollution rate must be in less well-protected industrial areas.

The problem of water pollution does not, of course, limit itself to fresh water. Massive oil spill disasters in recent years have made the public shockingly aware of the more general problem of the pollution of our oceans.

The eruption of the Ixtoc oil rig in the Gulf of Mexico on 3 June 1979 was certainly the worst single oil spill disaster to date. (Ixtoc was owned by the Mexican Pemex oil company which, five years later, would be responsible for the Mexico City propane disaster in which fifteen hundred people died.) The Ixtoc eruption created an oil slick some six hundred miles long and one hundred miles wide. It was nearly ten months before the well was capped, during which time hundreds of thousands of seabirds and mammals died by being coated in escaping oil. Furthermore, three million barrels of oil, worth at least $60 million, was lost.

The fact is, however, that even such massive oil spills as the Ixtoc disaster account for only a small part of the oil pollution of the oceans. Seepage from oil refineries in coastal ports alone accounts for 200,000 tons of oil being dumped into the sea each year. The total oil pollution of the oceans, much of it international, amounts to an estimated six million tons a year.

Human, agricultural and industrial sewage is a far more difficult problem to resolve than the whole issue of oil pollution. The oceans sustain major damage to their ecosystems through agricultural run-off because of heavy use of nitrate fertilizers and pesticides, and through industrial pollutants, especially chemicals like PCB's and heavy metals such as mercury, lead, cadmium and arsenic.

One of the most notorious sea pollution disasters became apparent in Japan in 1953. For many years the CHISSO chemical factory had been dumping mercury into the sea near the village of Minamata. First all the cats in Minamata died, then a large number of people appeared to be struck down with some unknown plague. The population of Minamata was dependent on fish as a main food source, and it was these fish, heavily contaminated with mercury, that were poisoning them. The results were frightening: paralysis of limbs, lack of muscular control, deafness, blindness, kidney failure, and brain damage.

Between 1955 and 1959, every third newborn child on Minamata suffered brain damage. It took years before the citizens of Minamata discovered the cause of their 'plague' – and even then criminal attempts were made to cover it up. By then 15,000 people were contaminated and over 100 people died of methyl mercury poisoning, which is now infamously known as 'Minamata disease'.

The degree to which the oceans are polluted may be gauged by the fact that penguins in the Antarctic and polar bears in the Arctic have been found to have heavy concentrations of DDT in their systems even though they are thousands of miles away from any source of pesticides. Dolphins and toothed whales have dangerous levels of PCB and mercury contamination, even though they spend all their life at sea. Large as the oceans of the world are, they cannot sustain infinite contamination – and many of their life-forms cannot sustain even very limited pollution.

As with the issue of fresh-water pollution, the main forum for attempts at resolving the international problem of ocean pollution has proved to be the United Nations. In 1974, the United Nations Environmental Programme instituted its Regional Seas Programme, with the objective of cleaning up major pollution areas in the world's seas and oceans. Its first initiative was its Mediterranean Action Plan which, in 1976, was signed by seventeen Mediterranean countries. While aware that a great deal more needs to be done to save the Mediterranean as a viable ecosystem, most environmentalists agree that the UNEP-inspired plan has been remarkable both in the way it has brought together the quarrelsome nations of the region, and in the degree to which those nations have co-operated on the issue of sea pollution. Since its inception UNEP has initiated a dozen such Regional Seas Programmes worldwide and is now supported in its efforts by thirty international organizations and over a hundred and twenty nations.

8

Gunfights at the Dump

EPA agent Alabama shootings and other incidents
• Organized crime in the dumps • Earth Day and
Decade of Environment 1970–80 • Reagan's EPA
'Sewergate' • Resignation of Ann Burford and
James Watt • Power of environmental lobby.

On 27 March 1983, three investigators for the Environmental Protection Agency in Atlanta, acting on a tip about an illegal toxic waste dump site, were searching the Alabama woodlands. They soon discovered the dump site: huge stacks of rusting and leaking fifty-five-gallon drums of chemical waste. As they began to investigate the degree of contamination on the site, gunfire suddenly erupted around them and they found themselves huddling behind the chemical drums for shelter. When the gunshots subsided, the unarmed agents fled the scene.

That same year, attempts to investigate toxic waste sites in Philadelphia resulted in numerous violent incidents in which federal agents were attacked by dogs, state inspectors were shot at, and one environmental inspector was badly beaten. Similar incidents occurred in Seattle, where a fire-bombing was perpetrated to hide a dump site, and in Chicago, where death threats silenced informers who tipped off EPA agents.

As the EPA itself points out, since 90% of all toxic waste is dumped illegally or unsafely, chemical garbage has become big business in the world of organized crime. There are tens of thousands of illegal dumps, many of which will result in leakages and poisonings at least as bad as that at Love Canal. Large criminal organizations are deeply involved in this illegal business and it is kept going by considerable political corruption.

Law enforcement officials and environmental agencies have an uphill struggle enforcing regulations, or even investigating sites in such an atmosphere. Nonetheless, in some places headway is being made. In 1984, Los Angeles boasted that its 'Toxic Waste Strike Force' had sent no less than twelve high-ranking company officials to prison for the illegal dumping of hazardous waste. The LA District Attorney claimed that he wished to deliver a clear message: 'Hazardous waste dumping is a violent crime against the community'.

In America the war against these environmental polluters has escalated since the advent of the first Earth Day in 1970, and the heightened public concern after numerous toxic waste-dumping scandals gained wide press coverage. In fact, the decade between the two Earth Day celebrations in 1970 and 1980 – when millions of Americans came out to 'celebrate the earth' and pledged to 'preserve, protect and clean up the planet' – has been called the 'Decade of the Environment'. The ten-year period certainly saw the most impressive and sweeping body of legislation enacted since the passage of a battery of civil rights legislation in the 1960s.

The National Environmental Policy Act of 1970 led to the creation of the vast Environmental Protection Agency whose function was to co-ordinate and manage federal pollution control programmes. The EPA was to enforce, among others, the impressive and newly amended Clean Air Act of 1970, and the Occupational Safety and Health Act, both of which dramatically improved air pollution standards in the United States during the seventies. The national standards for cleaning up the country's water resources were set by the Water Pollution Control Act of 1972 and the Safe Drinking Water Act of 1974. Two major legislative triumphs in the area of biological diversity were won with the passage of the Marine Mammal Protection Act of 1972, and the Endangered Species Act of 1973. Two further legislative acts won recognition for some of the more extreme problems presented by toxic waste when, in 1976, the Toxic Substances Control Act and the Resources Conservation and Recovery Act at last gave federal agencies control over toxic pollution.

In no sense, however, have industrial polluters disappeared. When President Carter introduced tougher dumping regulations, there was a huge increase in illegal 'midnight dumpings', and exporting toxic waste to the Third World became a sudden growth industry. The Carter administration vetoed such exports of toxic garbage, but the Reagan administration – ever eager to please industry – withdrew the ban.

Under Presidents Nixon, Ford and Carter, environmental legislation increased dramatically during the 'Decade of the Environment' and the EPA became a powerful, billion-dollar watchdog agency. Applauding the EPA and the environmental movement's achievements over the past decade when he opened the 1980 Earth Day celebrations, Jimmy Carter emphasized that the battle to save the environment was only just beginning. Sadly for the environmental movement in America, the following year saw the swearing in of President Ronald Reagan. The Reagan administration's handling of nearly all matters of the environment has been seen by most conservationists as regressive and even corrupt. Under Reagan's aegis, the Environmental Protection Agency and the Department of the Interior became packed with industrial lobbyists and lawyers whose chief function seemed to be to dismantle environmental legislation or to make it ineffective.

Evidence of extreme conflicts of interest and manipulation of funds led to what many journalists dubbed the 'Sewergate Scandal', which resulted in the resignation of EPA boss Ann Burford, whom Reagan had appointed. However, the greatest setback for Reagan's anti-environmentalists was the forced resignation of the powerful head of the Department of the Interior, the anti-ecology philosopher and hatchetman, James Watt. For many observers of the environmental movement in the United States the resignations of Burford and Watt – as powerful advocates of strongly regressive environmental policies – were seen as a major event in the balance of power in Washington. Philip Shabecoff of the *New York Times* saw them as heralding the political 'coming-of-age of the environmental movement'. The ability of the environmental movement to marshall public awareness and use intelligent legal and political manoeuvring made it a major political force that politicians could now ignore only at their peril.

9

Ecotage and Vigilantism

Ecotage: 'the Fox' of Illinois – anti-pollution Zorro •
Florida's Eco Command Force • Edward Abbey's
Monkey Wrench Gang and Earth First's 'sagebrush
terrorists' • Oregon air-raids • Guatamalan Army of
the Poor destroys 22 aircraft.

Environmental Action, the Washington-based anti-pollution group
that co-ordinated the first Earth Day celebration in 1970 to initiate a
decade of environmental activism, claims credit for coining the term
'ecotage' to describe non-violent ecological sabotage. However,
Environmental Action also credits the state of Illinois with giving the
world its first true 'ecotage commando'.

This rebel with a cause has never been unmasked, but during the
mid-1960s he was known as 'The Fox' of Kane County, Illinois. Mike
Royko, *Chicago Daily Express* columnist who was the Fox's media
contact, described him as 'an anti-pollution "Zorro" who has been
harrassing various companies, evading the police and making himself
a minor legend around Aurora and Kane County'.

The Fox's actions were motivated by outrage at industries that were
polluting and destroying the local countryside – particularly the local
Fox River. In his various escapades, and wearing many disguises, he
appeared in corporation offices and buildings to deliver the sludge and
dead fish that were the byproducts of their particular industries.
Elsewhere the Fox blocked factory sewage and drainage systems, and
sealed off smokestacks. Wherever he struck, he always left a note of
explanation, advising the company to 'clean up their act' and signing it
'The Fox'.

The ecotage actions were usually followed up by a poster campaign
against the offending corporation. During one such campaign, posters
and stickers attacking US Steel – one of the area's major polluters –
seemed to be everywhere. A sign sixty feet long appeared one morning
on a railway bridge over the main Indiana toll road. Playing on US

Steel's motto: 'We're involved,' the sign read: 'US steel: We're Involved – In Killing Lake Michigan'.

The Fox's bizarre career was an example to many environmentalists. In 1970, a Florida group calling itself the 'Eco-Command Force' planted yellow dye bombs in the sewage systems of eighteen Dade Country polluters in one night. As a result, half the county's canals turned bright yellow with the harmless dye. It proved a dramatic way of demonstrating to the citizens just how extensive the pollution in their region was.

Through the late 1970s and early 1980s environmental activists tried all manner of ecotage actions to draw attention to chemical pollutants. At one stage it seemed that every major industrial area of Europe and North America had a team of daring activists scaling its industrial smokestacks to hang banners of protest. One newspaperman suggested that smokestack-climbing should be nominated as a new Olympic sporting event.

The Wild West tradition in the western United States gave birth to an even more radical group of environmentalists called Earth First! This group of self-proclaimed 'sagebrush terrorists' openly supported ecotage operations that most conservation groups found unacceptable. Its symbol is a clenched green fist, and its formation was a classic example of how life can imitate art.

Earth First! is modelled on the activist philosophy of a fictional group called 'The Monkey Wrench Gang'. In his novel of the same title, the author Edward Abbey's fictional gang employs highly destructive eco-commando actions to fight industrial growth and pollution in the American wilderness. His gang practises wide-scale destruction of billboards, demolishes heavy industrial machinery and even plans to blow up a huge dam. Abbey's novel made him a godfather of ecotage in the west. In tribute to his ideas, the Earth First! newsletter advertises the sale of little silver monkey wrenches as 'a symbol for the discriminating eco-guerilla'.

Besides its newsletter, the group issues other useful literature on ecotage. One book, by leading Earth Firster Dave Foreman, is entitled: *Ecodefense: A Field Guide to Monkey Wrenching*. This instruction manual on ecological guerilla war tells the reader, in considerable detail, how to dismantle billboards, spike trees, sabotage roads, disable helicopters, burn bulldozers – and make a clean get-away.

Although none of the major environmental and conservationist groups wish to be associated with the Earth Firsters, the monkey wrench tactics of the group seem to have a strangely strong appeal to the normally arch-conservative rural working class in the western states who seem to bitterly resent the intrusion of large corporations in what they consider their wilderness regions. Earth Firsters, in fact, often portray mainstream environmental groups as hopelessly élitist and decadently upper-middle class. Sometimes calling themselves 'Rednecks For Wilderness', they argue that you need not have a doctorate degree in ecology to understand when a wilderness is being raped.

Yet, Earth First! support comes from a wide spectrum. Besides the central figure of author Edward Abbey, there is also Pulitzer-prize winning poet and early eco-guru, Gary Snyder. There are such individuals as George Crocker, who was involved in a Minnesota vigilante farmers' group called the 'Bolt Weevils' which became legendary for the spectacular destruction of sixteen high-voltage towers stretching massive power cables across their lands. Still another advocate of these tactics is *Sea Shepherd's* Paul Watson, who sees his organization as the sea-going equivalent of the land-based group.

One of the more unexpected supporters of Earth First! is David Brower, the architect and long-term director of the Sierra Club and founder of Friends of the Earth. Considering the animosity expressed by both organizations towards what they have labelled 'eco-terrorists', it may seem strange that the father-figure of both these groups now sides with the extremists. However, David Brower is one of the heroic figures in the struggle to save America's wilderness. For fifty years he has been in the vanguard of the environmentalist movement. 'These people are not the terrorists,' he says. 'The real terrorists are the polluters and despoilers of nature.'

Brower seems in agreement with Earth Firster Dave Foreman, who once wrote of direct action tactics: 'It's one tool. Sometimes you lobby; sometimes you write letters; sometimes you file lawsuits. And sometimes you monkey wrench.'

In the large forested areas of the American Pacific North-west, there have also been a considerable number of incidents of ecotage by citizens protesting over the issue of pesticide-spraying by giant timber companies and government forestry officials. In Keno, Oregon, in 1978, angry citizens blocked a public road and fired shots into the air in protest against a government spraying campaign. Elsewhere in

Oregon, spray vehicles had their tyres slashed, and on one occasion a spray helicopter was incinerated by a night raider.

Similar motivation in Index, Washington State, in 1980, resulted in thirty women and children throwing themselves across a railway track to stop a two-thousand-gallon chemical tanker from spraying railway right-of-way with chemical defoliants. When angry citizens in Oregon took action against a giant timber corporation believing they had evidence that the company's herbicide-spraying programme had caused a large number of pregnant women in one local community to miscarry, the company denied liability. Protesters became further enraged when one of the company's chemists attempted to minimize the importance of the issue with the statement: 'Babies are replaceable.'

Over the volatile issue of chemical spraying of populated areas, Eric Jansson of Friends of the Earth wrote a letter to the EPA that summed up the attitude of many irate citizens. 'I would like to suggest,' the letter read, 'that if it is legal to spray people with poison from aircraft and ground rigs without their permission, it should also be quite legal for anyone, including spray victims, to walk into your offices with pesticide cans and spray you with poisons.'

In some areas of the world, people have been provoked to far more extreme actions than those suggested by Jansson. At least a million people a year are seriously poisoned by pesticides and well over five thousand of these are killed. Among the worst offenders are the plantation owners of Guatemala who spray their million farm labourers as often as fifty times in the three-month harvesting and cultivating season. One clinic alone in a small Guatemalan town has treated up to forty severe pesticide-poisoning cases a day during the cotton-spraying season. Workers complained of vomiting, dizziness, delirium and damaged livers. Many became so severely ill that they died.

Under such conditions, it is not surprising that vigilante action is increasing. In 1976, in the town of La Flora, the 'Guatemalan Army of the Poor' launched an attack. Provoked beyond endurance, the vigilantes destroyed and burned twenty-two crop-dusting planes in a single night.

10

Scorched Earth Policy

Deforestation and desertification

'I will show you fear in a handful of dust.'
T. S. Eliot

1

The Cactus Rustlers

Case of the killer cactus • Cactus rustlers • Million
dollar black market • Arizona's cactus cops •
Cactus, drugs and illegal aliens • International
market for most endangered of all plant families.

In February 1982, David M. Grundman, a 27-year-old man from
Phoenix, went out into the Arizona desert and fired repeated shotgun
blasts into a giant saguaro cactus. The large pitchfork-shaped saguaro
is Arizona's state plant and a protected endangered species.
Widespread destruction of this cactus by collectors and vandals has
led state authorities to impose fines and even prison terms on
offenders.

However, enforcement has proved difficult and the saguaro is having
a hard time surviving. The giant cactus is not a prolific plant. It takes
twenty-five years to produce seed for regeneration and many cacti are
being dug up or destroyed long before they can reach this age.

The cactus Grundman fired on was twenty-seven feet tall, only about
half the maximum height the species is capable of achieving. Standing
a few feet away from it, Grundman concentrated his fire about four
feet up the base of the plant. When he let loose for the third time,
the trunk gave way.

To Grundman's surprise, the twenty-three foot upper section of the
cactus toppled directly on to him. Grundman tried to step out of the
way, but the cactus knocked the gunman to the ground with a vengeful
force, and perforated his body with thorns, for good measure.
Grundman was pronounced dead on arrival at Phoenix Hospital.

The 'Killer Cactus' episode in Arizona – one of the few incidents in
which plants have resorted to 'direct action' to protect themselves –
drew public attention to what has been a very one-sided battle in the
deserts of the United States.

The problem became so widespread that, in 1980, Arizona appointed
a seven-man force of 'cactus cops'. Investigations soon showed that
many of the offenders were part-time opportunists who combined a

desert picnic with a cactus round-up. However, a major part of the cactus-rustling business was conducted by some very tough professional outlaws.

As Richard Countryman, the first head of the cactus cops, pointed out: 'It's a rough bunch of guys we're playing with.' This proved to be a considerable understatement. Many of the outlaws are 'desert rats' who combine cactus-rustling with other pastimes like drug-smuggling and the transport of illegal aliens over the Mexican border. In several cases attempted cactus-rustling arrests resulted in exchanges of gunfire. In one case, a police witness was shot to death before he could testify in court against the alleged rustlers.

According to the botanist Lyman Benson, a leading cactus authority: 'The cactus family may be the most endangered species of all major plants.' However, the rarity of the cactus only serves to boost its price on the black market – which can be anything from $25 to $1,000 for each specimen – and has turned cactus-rustling into a multi-million-dollar illegal industry. America, Germany and Japan prove to be the major markets for these plants. In 1979, for instance, a single group of West Germans attempted to smuggle six thousand rare cactus specimens into Germany. In America, one dealer was found to be shipping up to 100,000 plants per month.

Beyond the crime of stripping the American deserts of their most distinctive form of plant life, ecologists point out that the removal of the cactus also destroys a significant part of the entire ecosystem of the desert. The cactus is the focal point of desert life, and the life-cycles of many other plant and animal species are dependent on its continued presence for their survival as well.

2

Orchid-Smugglers and Others

Smuggling of endangered orchids species • Giant
Sumatran orchids • US Venus Flytrap and Pitcher
Plants • 10,000 plant extinctions • Garrotting the
wine-palm and other atrocities.

The cactus is not the only plant that has become the victim of black-
marketeers and smuggling syndicates. As with the animal trade, there
seems to be a market for anything that is rare and exotic.

High on the list of commercially exploitable plants are orchids.
Thanks to worldwide air transport and refrigeration systems which
keep the plants fresh, extremely rare orchids can be cut from the
jungle and flown to markets around the world in a couple of days. As a
result, thousands of rare orchid species are rapidly achieving
endangered status in their jungle habitats of India, Latin America,
South-east Asia and the South Pacific. In India, at least ten species are
known to have become extinct in recent years through this trade, and a
score or more are certain to disappear in the next few years.

Many other flowering plants have also been seriously threatened by
the collectors of exotic species. One such plant is the astonishing
Sumatran *Rafflesia arnoldii*. A prima donna of the plant world, it
produces the world's largest flower which, at maximum size, measures
more than three feet in diameter. Also much in demand are the
'carnivorous' insect-eating species: the famous Venus Fly-trap of the
Carolinas – which was even sold as a mail-order item for a number of
years – and the Canebrake Pitcher-plant of Alabama.

As with the trade in exotic animals, few people have the knowledge
or sustained interest to keep exotic plants alive for long and about 99%
die within a few months of being removed from their habitat. Even if
some do survive, this does little for the species as a whole as their
disappearance from their natural habitat amounts to the same as their
being destroyed. They are unable to reproduce, and consequently they

may disappear entirely from their habitat in the wild – and eventually from the planet.

Unfortunately, the fate of extinction has befallen some five to ten thousand species of plants through man's depredations during the last three centuries. Many of these have been species that would have been extremely valuable to humans if they had been allowed to survive. The wine palm of Dominica was one such species. It was so named because of a fine clear wine that could be made from its pleasant juicy sap. The tree also produced a cherry-like fruit. Unfortunately, a process similar to garrotting was used to gather the sap, and this invariably killed the tree. Through ludicrous over-exploitation, the last wine palm was garrotted in 1926 and the species became extinct.

Many tree species of considerable commercial value have similarly been exterminated through short-sighted greed. The beautiful Mauritian ebony and the stately Juan Fernandez sandalwood are two notable trees that were made extinct through timber merchants felling every specimen they could find. Later, goats ate any seedlings that might have survived the clearing operations.

An estimated 20,000 vascular plant species have now been endangered by human exploitation. In many tropical forests and some island habitats, conditions are critical. The Hawaiian Islands have lost at least 300 unique species, and another 800 are endangered. Napoleon's place of exile – the tiny South Atlantic island of St Helena – had about a hundred endemic plant species: all but twenty of these are now extinct, and only five species are not critically endangered.

Tropical rain forests have a greater diversity of plant life than any other habitat, and those in the Amazon area have the greatest of all. It is estimated to have as many as six hundred plant species in each square mile. Unfortunately it is also the scene of the most widespread destruction of plant species in the world today. Many hundreds of thousands of these will become extinct even before they have been identified.

3

The Tree-Huggers

Ancient trees: redwoods, bristlecone pine, creosote
trees – 12,000 years old • Sacred woods • Flood
and drought through deforestation – millions die •
Chipko: Indian tree-hugging martyrs.

For some time it was believed that the giant sequoia redwoods of
California, being around three thousand years old, were the oldest
living things on the planet. They were indisputably the largest.
However, size does not necessarily prove age. In the mid-1950s it was
discovered that the small twisted desert trees known as bristlecomb
pines, that grew at the very limit of the tree-line of the White
Mountains of California, were nearly five thousand years old. More
recently, observations by scientists in the Arctic and the Antarctic have
led to the conclusion that the humble crustose lichens may survive for
ten thousand years or more.

In the 1980s the title of oldest living life-form on earth is held by the
desert creosote tree, which lives in the Mojave Desert. According to
Frank Vasek, the botanist who discovered the plant, it was probably
'one of the first life-forms to colonize the desert when the last glaciers
receded, and it has been a resident there ever since'. The scraggly bush-
like plant is estimated to be nearly twelve thousand years old.

Even before advanced botanical science made it possible to measure
the age of trees, mankind has equated them with longevity. Ancient
trees became emblematic of life itself and the tenacity of life in the face
of advancing time. Virtually every culture manifests some variation on
the motif of the 'sacred wood' in its mythology. The real and spiritual
life of man is often seen as being connected with the protection of
these sacred groves. In fact, myths serve as a very real lesson to us all.
We have every reason to fear the destruction of our forests.

As Earthscan pointed out in its global study, *Natural Disasters: Acts
of God or Acts of Man?*, six times as many people died in 'natural'
disasters in the 1970s as in the previous decade. (Interim reports

suggest that the 1980s will show an equally dramatic increase.) Yet there is no evidence that the natural forces behind these disasters, such as droughts, floods, cyclones or earthquakes have become more destructive or more frequent. What has changed is man's forestry, agricultural and urban policies.

Droughts, for instance, have greatly increased in number and severity. They are obviously the result of rapid deforestation and disastrous farming and herding activities. Forest destruction directly effects both rainfall and water-table levels, and it is estimated that one-quarter of the world's population is dependent, in one way or another, on water from tropical forests.

Floods – the flip-side of droughts – most often occur in the same areas as droughts but in different seasons. They too have been intensified by deforestation, overgrazing and the fatal consequences of soil erosion and desertification. Floods are, in fact, increasing faster than any other forms of disaster. In the 1960s floods destroyed or damaged the dwellings of over five million people; by the 1970s there were over fifteen million flood victims.

The mass deforestation of the Himalayan foothills is a classic case of forest destruction leading directly to massive flooding, disastrous soil erosion, crop destruction, mudslides, drying-up of natural springs and streams, and the widespread loss of flora and fauna. It has also led to the end of a way of life for the 'Hill People' who have always been dependent on the forests, forcing many to migrate into the already overpopulated agricultural plains.

In the face of this exploitation of forests, a grass-roots protest movement was formed in northern India in 1973. It is called 'Chipko Andolan' or the 'Hugging Movement'. 'Chipko', meaning 'embrace', had an historical precedent. In 1733, a girl embraced a tree to protect it from the maharajah's axemen who were about to cut down the forest near her village. Unfortunately, the action did not stop the axemen. The young girl was killed, along with 363 other village tree-huggers, before the maharajah relented and granted the villagers the right to maintain their own forest resources.

This tree-hugging tradition, a peculiar form of Gandhian non-violent protest, has spread widely since its resurrection in the state of Uttar Pradesh. The Chipko people know that their subsistence economy depends on the survival of the forests. Consequently, when timber companies attempt to open a new area to logging, the Chipko demonstrators – the majority of whom are women – literally throw themselves in front of the chainsaws and hug the trees.

In many instances the tree-huggers' tactics have paid off. A meeting between Chipko leader Sunderlal Bahuguna and Indira Gandhi resulted in a 1981 government ban on cutting trees above the 3,300-foot level in the Himalayas. Furthermore, the movement is spreading. Bahuguna has made a 3,000-mile march across the Himalayan range to organize local groups, while Chandi Prasad Bhatt – the movement's other leader – has initiated massive local tree-planting campaigns.

Wherever the Chipko organizers go they manage to spread a sophisticated message about forest ecology in their beguilingly simple slogan-riddle: 'What do the forests bear? Soil, water and pure air'.

4

The Green Revolution

UN Food Conferences 1974–84 • Green Revolution agriculture increases world's food production by one-third in ten years – yet 500 million starving • Billion poorest get poorer still • Without land and credit reform starvation will continue.

The United Nations convened its World Food Conference in Rome in 1974. At the time there was widespread crop failure. The Soviet Union and China were making massive purchases of grain from the West. There was famine in Africa and on the Indian subcontinent. It was estimated that 400 million people were starving or severely malnourished. Approximately 30,000 children a day were dying of starvation or hunger-related illness.

Everyone at the conference believed that a radical increase in food production was the obvious solution to the problem of hunger. As a result the UN committed itself to projects that would bring the 'Green Revolution' fully into the Third World.

The Green Revolution really began in North America and Western Europe in the 1940s with the development of a new generation of high-yield breeds of rice, maize and wheat which survived harsher climates, produced much higher volume and grew faster so that two or even three crops might be grown in a year. Its success in Western industrialized nations has brought about the most dramatic changes in the history of agriculture.

By the 1960s the Green Revolution had begun to spread to the Third World, but not rapidly enough. At the 1974 UN conference it was therefore decided to speed up this process so that the Third World might feed itself. There was an air of optimism for the future when Henry Kissinger spoke for many in attendance and pledged himself to the aim of the conference: 'Within a decade, no child will go to bed hungry . . . no family will fear for its next day's bread.'

By the time the United Nations held its 1984 World Food Conference in Brussels a decade later, the World Bank and many financial organizations the world over had pumped billions into agricultural programmes. The Green Revolution grains – which had so increased food production in the West – were beginning to achieve their production targets in the Third World. In the ten years since the previous Food Conference, world grain production had increased by a phenomenal one-third worldwide: from 1,200 million to 1,600 million tons annually. In many parts of Asia, Africa and South America grain production had more than doubled.

There was now more than enough food to feed the world. In fact there was a glut. Massive surpluses in Europe and North America cost the nations billions of dollars a year to stockpile because no buyers could be found. Farmers were paid handsomely not to grow food that would only add to the mountains that had accumulated. Agricultural production had exceeded even the best predictions in terms of overall levels.

Yet, for some reason, there were even worse famines in Africa and on the Indian subcontinent than ten years before. Instead of the 400 million starving or malnourished people, a decade of record food production had increased the number to 500 million. Instead of 30,000 children dying each day of starvation or hunger-related illnesses, there were now 40,000.

The world had more food than it could possibly use, yet more people were starving than at any time in history. The average African, for instance, was actually getting 20% less food than twenty years before. One fifth of the world's population – one billion people – live in conditions of 'absolute poverty' with no actual income. Another one billion live on an annual income of less than $400. In the midst of such plenty, how could this happen? How was it possible for the Green

Revolution to be such a success and yet such a failure at the same time. Unfortunately, in most of the Third World, the Green Revolution has simply widened the gap between rich and poor. Green Revolution grains grow most profitably on large-scale, highly mechanized farms and wealthy farmers are able to invest large amounts of money in fertilizers, irrigation, pesticides and sophisticated farm machinery. Consequently the small family farmer is pushed out of business by the large wealthy farmer. This is true even in America where family farmers are incredibly wealthy compared to their Third World equivalents. If the consequences of the Green Revolution in America have been devastating, in the Third World they have been little short of catastrophic.

Wealthy farmers are given massive financial aid to 'develop' modern industrial farms and huge dam projects or irrigation schemes are built by the governments to help these large-scale farming operations. The result is that those who are already well-off benefit the most, while conditions for the poor are actually made worse. Increased mechanization often means that the peasant-farmers are no longer even able to work for wealthy farmers at near-slave wages, let alone cultivate their own plots.

As the rich get richer the poor become poorer until they are pushed off the land and drift into marginal semi-desert lands or resort to slash-and-burn farms on unsuitable jungle terrain or on steep mountainsides. The result is the rapid erosion of land, the spreading of the desert, and the so-called 'natural disaster' cycle of floods, drought and famines in which millions die. It is an entirely predictable cycle, and the whole syndrome was unwittingly engineered by those who proposed the Green Revolution to such bodies as the UN as a means of solving the Third World hunger problem. This is not to say that the Green Revolution has been a total failure. In China, it has been an astonishing success. Whereas, before the Communist Revolution in 1949, millions of people routinely died of hunger each year, for the one-fifth of humanity that this nation represents, starvation has now been virtually eliminated. Fortunately for China's communist regime, the Green Revolution (though a decidedly lower-technology version of it) arrived at the same time as its political revolution. So, despite the fact that China's population has doubled since 1949, for the first time in its history all its people are being fed.

It seems that there is one basic reason for the success of the Green Revolution in China and its failure in the Third World – and it is a reason that need not be exclusive to communist regimes. Modern

China is committed to a system which maintains the peasant-farmer as the essential element in its agricultural policies. Its concern for the physical well-being of the average peasant-citizen is its main reason for attempting to increase crop yield and national wealth in the first place. Its leaders recognize that this human resource is the basic building block of the wealth of the nation.

In India – which possesses twice as much usable agricultural land per capita as China – Pakistan, Bangladesh, Indonesia, most of Africa and South America, this recognition is lacking. For the most part it is the poor peasant-farmer and the even poorer farm-labourer who are most in need of help and protection. Inevitably, it is exactly these groups – comprising about half of the world's population – that are given the least of either.

If, as stated, the aim of the United Nations was to feed the hungry of the world, the mechanics of achieving this should have been more carefully studied. Land reform and some system of credit to the truly poor are basic elements that have to be considered. Money-lending bodies and international development agencies which have funded, and continue to fund, the Green Revolution in the Third World are, at best, guilty of financing scandalously bad agricultural policies. At worst, they can be seen as the power behind the criminal exploitation of the poorest of the world's poor.

5

The Desert War

Live Aid and famine relief • Spreading deserts: 40% of Asian, 80% of African farmland at risk • One-third of landmass is arid: 700 million people • Famine belts of Africa and Indian subcontinent.

On 13 July 1985, the world's biggest pop concert was broadcast on television to an audience of over a billion viewers in a hundred and forty countries. The pop music industry has often been characterized as the most decadent form of business in a crassly capitalist society, but on that day, most of its major figures volunteered their skills to promote a totally altruistic and idealistic enterprise: raising money for the planet's most needy people.

Through the medium of television, the concept of the 'global village' was evoked to allow viewers all over the world to witness for themselves the great Ethiopian famine. Through a celebration of music, over $160 million was raised overnight to buy food for the hungry. It was achieved largely because one man, Bob Geldof, decided not to be stunned into inaction by the scale of the disaster. Instead, he organized Live Aid, the largest rock concert in history, and created a means by which – for once – the world's wealthiest billion people could individually contribute directly towards the welfare of the world's poorest billion. In both practical and symbolic terms Live Aid was a magnificent gesture and a wonderful event. Yet despite such monumental efforts by individuals, groups and agencies to relieve victims of famine and other disasters, actions of this kind can only be short-term emergency measures: they are simple first-aid tactics practised on victims of an ongoing war which mankind is waging with the spreading deserts of the world.

Unfortunately, the deserts seem to be winning the war. In 1888 an estimated 15% of usable farmland had been turned into desert or unproductive wasteland by human agricultural practices. Sixty years later, in 1948, that figure had doubled to 30%. Forty years after that, in 1988, it had doubled once again to a frightening 60%.

Intensive and unwise agricultural policies in even such wealthy nations as the United States have already resulted in desertification of croplands equivalent to an area greater than the entire state of California. In the Third World the problem is far more severe, yet the processes by which deserts are created continue unabated. Deforestation is followed by wasteful short-term agricultural practices. Large-scale wealthy farm-owners take possession of the most fertile lands and force poor peasant-farmers into cultivating submarginal lands. In many cases, the peasant-farmers also push out nomadic herding people who are forced to graze their animals on even more marginal land and so the deserts continue to grow. An estimated 40% of all arable land in Asia is considered at risk of desertification. In Africa, the figure is 80%.

One-third of the world's land is arid or semi-arid, and in these regions live 700 million people. They are among those most at risk of droughts, floods, famines, and epidemic diseases. The majority of these people live in the two great 'poverty belts' which stretch across the arid, semi-desert regions of central Africa and the Indian subcontinent.

The sudden call for famine relief in Ethiopia and the Sudan was not a surprise. The disaster was years in the making and on a predictably immense scale. Ethiopia and the Sudan are part of the poverty-belt lands of the Sub-Sahara. This region is called the Sahel and has experienced an ongoing series of famines since the early seventies. Ethiopia and the Sudan are only slightly different from Chad, Niger, Mali, Upper Volta and Mauritius in that their droughts, floods and famines have been somewhat more extreme, and have been aggravated by ruthless military conflicts and criminally negligent governments.

In 1977, the UN Conference on Desertification was called in Nairobi. Although it drew up a plan of action which was motivated by the disasters of the Sahel region, the famines of the eighties were far worse than the famines of the seventies, and the desert gained ground once again.

The Sahara, like many of the world's great deserts, is largely a man-made wasteland. The sands drifted rapidly northwards during the days of ancient Rome because the empire's demand for timber resulted in the destruction of all the great forests of North Africa. This deforestation was followed by a disastrous farming and herding policy which rapidly led to desertification in the naturally sandy soils of the region.

Similar strategies contributed to the Sahara's steady southward march over the same period of history and this continues even today. A similarly massive example of man-caused desertification is the Gobi Desert. This time it was the over-zealousness of the Chinese Empire which was responsible. The Gobi Desert covers 95% of what was, in ancient times, the fertile and highly productive Turfan Basin.

Today, there are major desertification problems in sixty-five nations, the worst of which are in the famine-belts of Africa and India. While the Sahara moves steadily south, the Rajasthan Desert in India is pushing its way east.

The former rain-forest lands of north-east Brazil – where one-third of the nation's population lives – are rapidly becoming desert. The Gobi Desert of China continues to expand at an alarming rate. In the Soviet Union the Aral and Caspian Seas are drying up and reverting to salt flats. Vast regions around them are becoming desert wastelands.

6

Men of the Trees

'Live fences' of Egypt • Paved dunes of Libya •
Green Wall of China • St Barbe Baker's Men of the
Trees movement • New Supertrees and plants:
'shock troops' against desert.

There are those who fight the desert's advances head on. Israel's systematic pushing back of the desert with irrigation schemes and tree-planting is well documented. Egypt and such places as Upper Volta are attempting to halt, and even turn back, desert expansion by planting 'live fences': hedges of tough desert trees.

Oil-rich Libya has come up with a bizarre scheme to stop expansion by paving the desert. It is a scheme of planting eucalyptus and acacia trees on dunes, and then spraying the entire dune with a layer of asphalt in order to stop it shifting and burying the newly planted trees.

One of the grandest schemes is taking place in China where a modern Great Wall is in progress. This is the Green Wall of China: a wall of living trees which will, in fact, be planted roughly parallel to the ancient stone Great Wall. The Green Wall is designed to protect the nation from the invasion of the cold winds and sandstorms out of Siberia and Mongolia that have helped to create another 25,000 square miles of desert in China in the last forty years.

The Green Wall is part of a massive reforestation scheme. Since 1950, 100,000 square miles of trees have been planted, and the current plan is to plant another 200,000 square miles of trees between 1980 and the year 2000. This planting scheme would make 20% of China forestland, and further schemes are expected to result in a 30% figure before 2050.

Those who are fighting to control desert expansion by reforestation are also becoming aware of the complex relationship between rain and non-coniferous forests. China lost 50% of its annual rainfall this century through deforestation. It is hoped that once forests are regrown the rain will return. This has certainly proved to be the case in other parts of the world.

The father of the grand-scale reforestation programme is Dr Richard St Barbe Baker. Since the end of the First World War St Barbe Baker has spent his life reforesting the planet. In 1920 he founded 'Men of the Trees' in Kenya. Enlisting the help of nine-thousand voluntary tree-planters he attempted to arrest the advancing desert in Kenya. Since then, St Barbe Baker has started Men of the Trees movements throughout Africa, the Near East, India and New Zealand, amongst other countries.

In North America, St Barbe Baker launched a Save the Redwood Fund, and during the depression years under President F. D. Roosevelt, he organized a plan which resulted in a 1,000-mile-long shelter belt of trees across America. It was a scheme which employed some six million workers and did a great deal to reclaim the 'Dustbowl' prairie lands of America.

Today, a great deal of research is going on to determine which trees are the best and most viable in any given region. With tropical timber being felled at an annual rate equivalent to the land area of Cuba, the speed of reforestation is a major consideration. Also, it is often pointed out by many Third World politicians that the greatest fuel crisis today is not oil or coal, but – for most of the world's population – firewood. Consequently, considerable emphasis has been placed on the development of so-called 'supertrees'. These are drought-resistant, rapid-growing trees which have been called the 'shock troops of the war on deforestation'.

Most of the supertrees are legumes: species of trees related to lima beans, lentils and chickpeas. Until recently, they have been largely ignored by industry but this is no longer the case in view of their extraordinary diversity of use and their ability to produce wood at a rate ten times greater than other trees, in extremely poor soil.

The most impressive of the supertrees must be the leucaena of South America and South-east Asia. There are over a hundred varieties of leucaena, ranging from sixty-five foot trees suitable for timber to fifteen-foot hedge trees. They are drought resistant with edible foliage and many can grow up to twenty feet a year. A favourite is the *Leucaena leucocephala*. The largest of these trees, it can grow to a height of sixty-five feet and to a thickness of sixteen inches in five years. It is a good source of building timber, furniture wood, and paper. It can also be utilized as firewood, charcoal and a kind of natural gas. Its leaves have multiple uses as nitrogen-rich fertilizer, protein-rich cattlefeed, or sweet protein-rich food for humans. Nor is this all, for the tree's seed pods can be eaten raw, cooked, or ground into flour and baked. The gum of the tree can be used as a thickener for foods and cosmetics, and even the tree's white flowers serve a purpose in being popular with honey-producing bees.

Other species of rapid-growing 'shock troop' trees are the acacia and gmelina trees which are speedy producers of fuel wood and pulp for paper manufacturing. The Amazonian copaifera is another rapid-growing tree with a unique sap which can be used to fuel diesel engines, while the crimson caliandra has a natural fire-retardant that can make the tree an excellent natural firebreak hedge to stop the spread of brush fires.

Nor are trees the only arid land plants used in this war against the desert. Many shrubs and plants are being planted to hold the soil against the spreading deserts. Among them are the jojoba shrub which produces a fine, high quality oil from its seeds. Another Mexican desert shrub, the guayule, produces a natural rubber with many commercial uses. The fast-growing ground vine called the buffalo gourd produces a serviceable cooking oil, while the tough mesquite plant has edible pods that have a high protein and high sucrose content.

Other arid land plants with possibilities are the mongongo nut and the tsi bean – two of the most protein-rich plants in the world. They thrive in semi-deserts, but are currently eaten only by the San tribal people of Botswana and Namibia.

It has often been observed that 95% of human food comes from thirty plants, and – even more remarkably that 75% comes from only eight of these plants, which is an extremely narrow dietary basis for the species. Such epidemics as Dutch elm disease have demonstrated the near impossibility of stopping plant plagues once they take hold. If a major epidemic struck any of the cereal crops – such as wheat, rice and maize – there would be death by famine on a cataclysmic scale.

Awareness of how dependent we are on a very few plants has, in recent years, emphasized the need for the introduction of new wild plant species and the establishment of plant gene banks for the creation of new grains and maizes, fruits and trees.

It was the great American Indian horticultural civilizations of Mexico and South America that domesticated many wild species and gave us maize, potatoes, beans and numerous other food crops. However, many other root crops and varieties of corn and bean which were thriving in the new world at the time of European contact did not continue in cultivation. Many of these are now being hunted down in remnant populations and are proving stronger – and more disease resistant – than currently farmed species. Also, wild species of plants

are being interbred with domesticated species in order to enhance
their food value and hardiness.

The vast rain forests of Latin America are widely recognized as the
richest lands in the world for plant and animal species. It is these lands
that hold the key to the thousands of food, pharmaceutical, and
industrial products that are still to be discovered – but only if we don't
entirely destroy the massive genetic reserves that the rain forests
represent. Many scientists warn us that, by turning jungle land into
desert, we may very well be destroying the genetic basis by which
mankind might sustain itself in the future.

7

Chainsaw Massacres

**Vietnam's anti-forester rangers • Agent Orange •
Destruction of Amazonia • Protectors of the forests:
the 200 million tribal people of the Fourth World •
Blowpipe War • Sarawak • Genocide in Amazonia
• 1988 Tikuna Massacre • Martyrdom of Chico
Mendes.**

Jungle warfare took on a new meaning during the Vietnam conflict
when the Americans decided to wage war on the jungle itself. A
division of 'Anti-Forest Rangers' was established to eliminate vast
tracts of forestland used by Viet Cong troops with a lethal combination
of napalm and a chemical defoliant called Agent Orange. These Anti-
Forest Rangers used a parody of Smokey the Bear's fire prevention
message as their corps motto: 'Only *we* can prevent forests.'

However, most of the war against the jungles of the world goes on in
a less flamboyant manner than the 'scorched earth' tactics that were
practised in Vietnam.

The 'environmental activists' who stand most determinedly against
the logging companies and slash-and-burn peasant-farmers are people

who for the most part have never heard the word 'ecology' and yet know better than anyone how their lives and the life of the forest are directly linked. These are the tribal peoples of the so-called Fourth World: wilderness and forest-dwelling people who do not need the explanations of science to understand that if tropical forests are allowed to vanish, they will vanish with them.

These people who live on all the continents of the world, and are as directly endangered as the forests themselves, have put up a brave but usually hopeless fight against bulldozers and aircraft. They often live in small tribal groups, but in all they make up over 200 million people – a scattered worldwide 'nation', with a population roughly equal to that of America.

Such tribal groups in the Solomon Islands, for instance, rose up in armed rebellion against ruthless timber interests. Similarly, in Borneo, the Penan and Kayan tribesmen of Sarawak, armed with spears and blowpipes, have blockaded logging operations backed by wealthy Malaysians. Typically, concessions covering most of Sarawak's forests have been given to timber firms with direct or indirect links with corrupt government officials. Among the combatants in this conflict is the strange figure of Bruno Manser, a Swiss anthropologist who has 'gone native', and lives as a white tribesman among the Penan. This Robin Hood of Sarawak is especially resented by authorities who have placed a bounty on his head.

In the Philippines, similar abominations have been committed, with tribal people being massacred or driven out of their forest homelands. In Thailand, not only do timber companies use guns against the forest people, but there are pirate companies with their own armies who will stop at nothing to fell timber in parklands and other illegal areas, including hijacking freight trains and trucks to ship it out of the country.

By November 1988 it was obvious the pirate loggers of Thailand had gone too far. The worst predictions of the conservationists did not compare with the destruction brought on by heedless deforestation.

Well over a thousand people are estimated to have died when whole logged-off mountain sides slid down and buried entire villages. The terrible mudslides and floods, triggered by heavy November rains, resulted in tens of thousands being left homeless. Scores of schools, hospitals, government offices, roads, railways and bridges were swept away. Weeks after the initial floods, hundreds of people were being buried in further mudslides bringing down boulders, trees and logs. Damage was estimated at hundreds of millions of dollars – in strict financial terms, many times more than was made from cutting the forests in the first place.

The Prime Minister, Chatichai Chonnavan, was shocked into acknowledging the reasons for the disasters and called for a complete ban on all logging in Thailand. 'This serves as an expensive lesson for all of us on the adverse effects of excessive logging'.

Others put it more bluntly claiming the logging companies and their allies ought to be charged with mass murder.

Despite the excesses of other regions, the most critical forest battleground must be that of Amazonia. The Amazonian rain forest is the size of the whole of Europe and the tribal peoples there are making a last stand. Once they boasted a population of four to six million people. Nearly one hundred tribes have become extinct in this century alone. Today, only two hundred thousand forest people survive.

On the morning of 28 March 1988, a group of ninety unarmed Tikuna men, women and children made a canoe journey to Capacete in the extreme west of the Brazilian Amazon to meet an Indian agency lawyer and a police officer to register a complaint about a timber operator. The officials did not turn up. Instead, the timber operator appeared with twenty armed men. On arrival the gunmen simply opened fire on the unarmed Indians. They shot men, women and children indiscriminately. At least fourteen were killed, and twenty-seven others were eventually taken to hospital with severe gunshot wounds.

This massacre is only one of the more recent to reach the Western news media. Amazonia has become a horrific and bloody 'Wild West Show' fought with automatic weapons and aircraft. The tribal people are being eliminated in order to take their land from them.

There is a certain consistency in Indian history in Latin America during the last four centuries. It is a chronicle of systematic slaughter of its millions of tribal peoples, most of which has been pointedly ignored by the rest of the world. There are now considerably fewer Indians left to exploit, but during the past thirty years, there have been literally thousands of incidents in which tribal peoples were bombed, shot and poisoned by soldiers, mercenary killers, ranchers, farmers, timber merchants and prospectors. Even missionaries and Indian agents have been involved in organizing armed manhunts to kill or enslave Indians. Young women and boys are abducted and sold as slaves and child prostitutes. Others who survive direct conflict suffer only slightly less extremely. Once they have their land taken from them, they usually end up in a slightly different form of slavery called debt-bondage.

However, the most destructive factor in the lives of these people is the new germs which foreigners are bringing to their jungles. Such simple childhood diseases as measles and chickenpox prove as fatal as the scores of more dangerous introduced diseases to a people who have no resistance to them.

When roads are built through areas of virgin forest, Indian settlements often lose a quarter to a half of their populations to disease in a single year. On many occasions these diseases have been intentionally spread among the Indians, often by Indian agents, or others who have been sent into the jungles supposedly to protect them.

As we have seen, in recent years the Indians have made courageous attempts to fight back but, by most South American laws, they have virtually no legal rights. A few organizations like the London-based Survival International – which lobbies on behalf of tribal peoples worldwide – have taken up their cause. However, Fourth World people are typically placed outside the framework of the world's political and legal structures. They are often granted less legal protection than many species of endangered animals.

Still, against all odds, some of these forest guardians have won victories in legal and land rights. Although not seen as such by the governments who deal with them, these tribal peoples are winning victories on behalf of peoples of the world. And, although each year the Indians see more of the forest being relentlessly pushed back, they are now winning allies within the Amazon as well.

One such ally was Francisco (Chico) Mendes, the Amazon rubber tappers' leader who, on the 22 December 1988, was assassinated in the Amazonian town of Xapuri. Chico Mendes had attempted to organize resistance to the powerful interests who were rapidly destroying Amazonia's forest resources. Their response was to hire professional killers to eliminate him.

So far as the forests were concerned, 1988 was the worst year in Amazonia's history. An area the size of Holland was burnt off by cattle ranchers and settlers. But 22 December 1988 may prove to be an historic day: Mendes' assassination has become a focal point in the fight to save Amazonia. However it must be said that this killing was a part of a long-standing range war. The Cattle Barons and wealthy land-owners, who make up 1% of the population and control 50% of the land, have used hired killers to assassinate any who attempt to advocate agrarian reform. In the two years leading up to Mendes' killing, hired 'pistoleiros' committed over 700 contract murders. Of these only two pistoleiros – both priest-killers – have ever been tracked down and convicted.

Mendes has rapidly become a martyr for groups worldwide who fight for the preservation of the rain forests. Immediately after his murder, Mendes was the subject of an angry editorial in the *New York Times* entitled, 'Brazil Burns the Future'.   'If Brazil wants the world's sympathy on matters of debt or democracy, it cannot ignore the international outrage at assaults on the environment and those who defend it,' stated the *Times*.

Indeed, the recent blocking of a $500 million World Bank loan by a disapproving foreign community has delivered an unmistakable message to Brazilian politicians that the ghost of Chico Mendes was a far more formidable opponent than the rubber tapper from Xapuri had ever been while alive. As a spokesman for the Environmental Defence Fund in Washington DC stated: 'Mendes taught a valuable lesson: social justice and environmental protection in the Amazon are inseparable.'

8

Acid Rain

Anti-green politics: Reagan's tree-pollution theory • Industrial pollution and acid rain • Destruction of forests • Major acid conflicts in Europe and North America.

President Reagan achieved instant fame in environmental circles early in his political career in California when he supported the expansion of logging operations, and opposed further protection of the giant redwoods. 'After all,' he claimed, 'a tree's a tree. If you've seen one giant redwood you've seen them all. What's the big deal?'

With this kind of record, environmentalists did not expect an entirely sympathetic attitude to be forthcoming from the Reagan administration. However, they were not prepared for the extraordinary hostility – and what could only be described as a determinedly ignorant approach of the White House – towards issues of ecology.

In one of his weekly presidential radio broadcasts, Ronald Reagan addressed the nation on the issues of forestry and air pollution. The President decided to set the nation straight on a few basic points of history and science. He pointed out firstly (and entirely wrongly) that there were now more trees and forestland in America than at any time in history. Secondly he warned that we should not be overly protective towards our forests because it was an undisputed fact (wrong again) that 'approximately 80% of our air pollution stems from hydrocarbons released by vegetation'.

Shortly after this broadcast an ecologist came forward with a request to demonstrate the relative safety of trees compared to that of an internal combustion engine. If he offered to be locked in a garage for 24 hours with the most dangerously transpiring tree to be found, would Reagan offer to demonstrate his point by doing the same with a Ford pickup with the engine running?

If Reagan's theories on ecology proved a source of amusement to some, to others they were a source of total despair. How could politicians be expected to come to terms with the often complex issues of ecology, if the leader of the world's most powerful nation could not even grasp scientific theories – like photosynthesis – that an average elementary school student has no difficulty comprehending? As a general rule, however, politicians will only understand those environmental theories and ideas that it is politically or economically expedient for them to understand.

Acid rain is one of the most volatile environmental issues of the decade. It is an issue that has had wide public exposure and the destruction it has caused in the nations of the northern hemisphere is extraordinary. Yet both the governments of Britain and America – despite the howls of outrage from their neighbours – claim that the case against air pollution and acid rain has not been proved to them. There is a logic to this sort of defiant stupidity. It stems from the fact that both America and Britain are massive exporters of airborne pollution and acid rain, and both wish to avoid the financial burden of cleaning up their industries. They would rather allow their industries to continue to profit by dumping their pollution on their neighbours.

Acid rain, sleet, snow, fog and even dust are created when heavy concentrations of sulphur and nitrogen oxides enter the atmosphere. The major source of these oxides is the huge and familiar smokestacks of fossil-fuel-burning plants, industrial factories and smelters; a secondary source is the exhaust fumes of automobiles. When gases containing sulphur and nitrogen oxides are released into the air, prevailing winds often carry them hundreds of miles away from the source of pollution before they come down to land. In the process they are transformed into diluted forms of sulphuric and nitric acids. Some rainfall in heavily polluted areas is equal to the acidity of lemon juice or vinegar. In the long term acid rain destroys lakes, forests, crops, groundwater and even corrodes stone and steel.

One has to sympathize with victim nations like Canada, Norway and Sweden. Although there are major concentrations of industry in southern Ontario and Quebec, about 50% of Canada's acid rain pollution is imported from America. Some 10,000 lakes are so acidified that they cannot support fish, and Canada's national tree, the sugar maple, is being devastated. In Norway and Sweden where over 20,000 lakes are being killed off, 80%–90% of the pollution drifts in from Britain and the rest of Europe. In Germany a third of the Black Forest is dead or dying, and acid rain damage is costing the timber and agriculture industries about $1.5 billion each year.

The prevailing winds that blow northward, over the industrial belts of Europe and America, are creating massive scorched-earth corridors through the lands of their neighbours. The winds are filled with acidic poisons that literally burn the life out of all in their path.

It is the conservation-minded Scandinavians who have most prominently promoted reform in the area of acid rain. They point out that acid rain is an international problem that can only be resolved by international co-operation. To this end, Sweden has promoted the 30% Club. In Ottawa in 1984 a score of nations signed an agreement which committed each to reducing its sulphur dioxide emissions by 30% within a decade. Predictably, both the United States and Britain refused to sign on the grounds that there was insufficient evidence for acid rain, promising only further research.

9

The Greenhouse Effect

Doomsday theory: carbon dioxide build-up, temperature rise and melting of polar caps • Worldwide flooding • Ozone layer depletion • Cancer-inducing radiation • CFC's and aerosol spray control • Atmospheric management linked to forest management • Forests equal 75% of planet's biomass.

Beyond the many arguments put forward for saving the great forests of the world, a new theory concerning the ultimate consequence of their destruction has recently arisen. This is a rather startling doomsday theory of global dimensions.

The theory relates to the planet's atmosphere. The problem is that the current rate of massive forest destruction is resulting in less carbon dioxide being converted into oxygen. And, regardless of Ronald Reagan's views on the matter, the forests of the world are the major oxygen producers of the planet. The rain forests of Amazonia, for instance, generate up to one-quarter of the earth's oxygen – earning it the title of 'the lung of the planet'.

At the same time as oxygen-producing forests are being reduced, we have radically increased the burning of fossil fuels and other means of adding pollutants to the atmosphere, with the result that carbon dioxide levels have risen by 25% in the last century. Projections into the future suggest that these levels will increase by a further 100% over the next century.

If this should occur, it is predicted that the so-called 'Greenhouse Effect' will come into operation. Although greenhouses work on a slightly different principle, the idea is that increased carbon dioxide brought about by air pollution and forest destruction will result in a heating-up of the earth's temperatures. This shift in the overall world temperature would result in a dramatic shift in climates. It is believed that North America's grain belt, for instance, would suffer from permanent drought and become a vast desert. Other major agricultural areas would become either too wet or too dry for crops. The most dramatic effect, however, would probably be the melting of sections of the polar ice-caps. It is thought that the release of these huge reservoirs of frozen water would result in the planet's sea-levels rising by ten to twenty feet. This would cause massive flooding that would put the cities and lands of nearly half the world's population under water.

There is considerable debate about the Greenhouse Effect and the related consequences, as there is with all air pollution concepts, because of the conflicting interests of various governments, industries and public interest groups. Carbon dioxide build-up is one such issue, sulphur and nitrogen oxides causing acid rain are another, and a further specialized issue concerns the release of pollutants which cause the destruction of the planet's ozone layer.

Ozone is one of the planet's so-called 'greenhouse gases' which is found in a layer of the stratosphere about ten miles above the earth's surface. This layer has the beneficial effect of filtering ultraviolet radiation from the sun's rays. One form of radiation, UV-B is particularly harmful and is the cause of cataracts and skin cancer. A US Environmental Protection Agency study estimated that nearly a million additional cancer deaths over the next century may result from the depletion of the ozone layer.

Serious attention was only given to ozone depletion when satellite photographs clearly revealed a massive 'hole' in the ozone layer over the South Pole. Subsequent studies revealed that ozone over the pole had depleted by 40% since the 1960s and the hole was rapidly growing greater each year.

One of the major causes of the depletion of the ozone layer is the release of CFCs (chlorofluorocarbons), which are largely created by the chlorine content in aerosol spray propellants and refrigerator coolants. Other gases causing serious disturbances in the ozone layer are nitrous oxide and methane, the by-products of industry and agriculture.

The most immediate concern has been focused on CFCs, as these gases have a pollution capacity 10,000 times greater than carbon dioxide, and even released in moderate amounts cause extreme destruction in the ozone layer. Although initially reluctant to take steps to control the powerful chemical industries, governments have been gradually persuaded to do so. In 1987 an international protocol was established to commit all signatory nations to reducing CFCs by 35% by the year 2000. Many critics and citizens felt that this was too little, too late. Some of these criticisms seem to have reached the right ears, however. In 1988, Du Pont, the American chemical giant and the world's largest producer of CFCs, promised a 100% phase-out of CFCs within twenty years. Many are now hopeful that others will follow suit.

Dr Joseph Farman, the scientist on the British Antarctic Survey who was the first to link CFCs with the polar ozone hole, welcomed the recognition of the problem by industry and government. However, he wished to point out: 'The real crime has been that the chemical companies and the governments have set a heavy burden of proof on the environmentalists.' Certainly the motivation for responsible controls seldom, if ever, comes from those within industry or government.

Dr Farman continued: 'Sooner or later we must force ourselves to manage the entire atmosphere. CFCs are just one danger. Other emissions, like carbon dioxide, methane and nitrous oxide all demand attention.' He is certainly right. Although there has been considerable resistance by industry, CFCs are among the easiest of the pollutants to control because of their limited – if widespread – use. As ongoing research into acid rain and the Greenhouse Effect is demonstrating, other pollution factors are far more difficult to bring under control.

Some believe that the Greenhouse Effect is already measurably having its influence on the planet. Scientists during the 1980s have registered the hottest annual temperatures since records have been kept. Although it may be too early to be sure, some believe that the earth's atmosphere is already overheating, and the shifting rainfall patterns and droughts in Africa, India and America are seen as symptomatic of this global increase in temperature.

Whether we are now experiencing the beginnings of the Greenhouse Effect or not, it is becoming increasingly clear that we must now take steps, as Dr Farman said, 'to manage the entire atmosphere'. In June 1988, representatives from six continents gathered in Toronto specifically to address these issues. It was the first concerted international effort to push for the implementation of an International Law of the Atmosphere to parallel the Law of the Sea.

Control of airborne pollutants is a major concern in any such plans for management of the atmosphere. However, as the Toronto conference recognized, control alone is not enough. Any strategy must include the conservation and management of the world's great forests as the planet's major 'oxygen-factories'. After all, photosynthesis was the means by which the earth's atmosphere first came into being, and without plant-life continuing this process, the atmosphere, as we know it, would simply cease to exist.

The worldwide consequences of forest destruction should not be entirely surprising to those who ponder the issue for long. After all, forests provide the earth with no less than three-quarters of its total biomass. Forests are *the* resource base of the planet.

11

Anti-Nuclear War

Perspectives on the anti-nuclear movement

'The unleashing power of the atom has changed everything except our way of thinking . . . we need an essentially new way of thinking if mankind is to survive.'
Albert Einstein

1

Sinking the *Rainbow Warrior*

Bombing of Greenpeace ship, 1985 • Murder of
Fernando Pereira • Muroroa nuclear protests •
French agents and the 'Underwatergate' affair •
Government resignations and diplomatic
blackmail.

On 10 July 1985, the Greenpeace flagship, *Rainbow Warrior*, was
berthed in Auckland Harbour, New Zealand. It was the eve of her
voyage into the French nuclear testing site of Muroroa Island. The
Greenpeace ship had already made a voyage of protest by sailing
through the Marshall Islands where America had tested 66 of her
nuclear bombs thirty years before. There the *Warrior* crew had helped
evacuate the native islanders of Rongelap Atoll who had for years
appealed to the authorities to take them off their contaminated island.

Greenpeace documented the numerous reports of contamination,
birth defects and cancers. They photographed the people and their
move to the 'safe' island of Mejato. Then they sailed to New Zealand to
support Prime Minister David Lange's call for a 'Nuclear Free Pacific'.
From here the *Warrior* was set to lead a flotilla from New Zealand to
Muroroa Island.

At ten minutes to midnight, Auckland harbour was rocked by an
explosion. Beneath the harbour lights, the 150-foot, 400-ton *Rainbow
Warrior* lurched and began to keel on her side as the waters rushed
into her engine-room.

When the explosion struck, most of the crew members ran in
darkness for the upper deck. Davey Edwards, the British first engineer,
rushed down to the engine-room to find that the engines were already
submerged. Captain Peter Wilcox, quickly realizing the seriousness of
the damage, gave orders to abandon ship but the Dutch first mate,
Martini Gotje, and the Greenpeace photographer, Fernando Pereira,
had already gone below; Gotje to check cabins for missing crew, and
Pereira to salvage what he could of the photography equipment.

Suddenly there was a brilliant flash in the water beneath the *Warrior*. The ship jolted even more violently. A second massive explosion punched a truck-sized hole in the steel hull and the thunderclap of the blast rolled across Auckland.

Martini Gotje was climbing back up the stairs and Davey Edwards was pulling the elderly relief-cook, Margaret Mills, from her lower-deck cabin when the second explosion battered the ship. Gotje leapt for the upper deck, and Edwards and Mills fled with all the speed and strength they could as the seas raced down the corridors with them. They reached the stairs with the water swirling about them and as they climbed, so did the water. Somehow, Gotje, Mrs Mills – still in her soaking pyjamas – and Davey Edwards made the upper deck and scrambled on to the wharf with the others.

It was all over. Four minutes after the first explosion, twelve crew members were ashore and the *Rainbow Warrior* had completely flooded and keeled over. She was only partly above water because she was hanging by her moorings.

It would be another two hours before navy divers would be able to recover the body of the thirteenth Greenpeace crew member. This was the Portuguese-born Hollander, Fernando Pereira. The Greenpeace photographer was thirty-three years old and left behind a wife and two young children.

The Greenpeace conflict with France over Muroroa Island began in 1972 when its current international director, David McTaggart, sailed a protest vessel into the test area for the first time. His yacht was rammed by a French navy vessel and seized. In 1973 when he returned, his yacht was boarded, this time by French commandos, who savagely beat McTaggart and another crew member before again seizing his boat.

The Greenpeace protest had been successful in helping to draw media attention to the test activities and was instrumental in bringing atmospheric nuclear testing to an end in Muroroa. However, underground tests continued and, after numerous leaks and contaminations, culminated in an incident on 6 July 1979 when two men were killed and six seriously injured during a decontamination operation in an underground bunker. Acetone vapour exploded in the accident and, as a result, a cloud of plutonium was released over the island.

After the bombing of the *Rainbow Warrior*, it was fortunate for Greenpeace that the New Zealand government was firmly supportive of their protest effort. Much to the anger of the French, Americans and British, New Zealand has pursued a Nuclear-Free South Pacific policy. Prime Minister David Lange issued an early statement that his government was treating the action as a high-priority terrorist act, resulting in homicide.

It was fairly obvious whose interests were served with the sinking of the *Rainbow Warrior*; however, it was still a startling revelation when it was discovered that the massive twenty kilos of explosives used to blow up the ship had been planted by commandos from France's secret service – the DGSE Action Service.

The scandal that followed in the wake of the arrest of Major Alain Mafart and Captain Dominique Prieur, two of the seven agents involved, soon brought about a political crisis in France. French government sources immediately and vehemently denied it had ordered the action and politicians of all parties closed ranks in a spate of absurd nationalism in an attempt to cover up the scandal. However, the French press succeeded, despite political and legal intimidation, in exposing the details of what was inevitably labelled France's 'Underwatergate'.

Despite government attempts to cover up the incident, the press ensured that at least part of the truth emerged. The revelation that the operation was probably approved at cabinet level came close to bringing down the government. It was only saved by the sacrifice of both the chief of the DGSE secret service, Admiral Pierre Lacoste, and the Minister of Defence, Charles Hernu.

Before the year was out, the two captive agents had pleaded guilty in a New Zealand court on a charge of manslaughter. This prevented further prying into the affair which would come with an open trial. However, the silent agents were each given ten-year prison sentences for the killing.

Although now admitting responsibility for what was, after all, an act of terrorism and murder in a country that is considered a friendly ally, the French were outraged with what they considered extreme sentences for their agents. Prime Minister Chirac stated openly that the New Zealand Prime Minister would not be welcomed in France until the day that these agents were released. Largely because of the Lange government's anti-nuclear policies, leaders in America, Britain, and the rest of Europe did not choose to pass comment on this action of state-terrorism against New Zealand. Instead they stood back and watched France apply a shameful and crippling barrage of trade sanctions on the economically weaker New Zealand. Eventually, by a combination of trade sanctions, sabotage by French customs officials, diplomatic intrigue, and economic blackmail, France forced the reluctant New Zealand to release the agents within a year of conviction.

In the end, France did agree to pay substantial compensation to New Zealand, Greenpeace, and the family of Fernando Pereira. The settlement was conditional on the convicted agents being placed within the 'custody' of the French authorities on a relatively remote French colonial island where they might serve out a minimum of three years in exile before being permitted to return to Europe. However, still angered at the audacity of New Zealand daring to prosecute its agents in the first place, and smarting from the necessity of having to make financial reparation, the French government could not resist contemptuously thumbing its nose at the Lange government again over the incident. Once its agents were safely in French territory, the terms of the agreement governing custody, travel restrictions and exile were promptly ignored, and Prieur and Mafart enjoyed a certain celebrity status in France as martyred patriots.

2

Killing of Silkwood

**Mystery of Karen Silkwood's death, 1974 •
Contamination at Kerr-Magee nuclear plant •
Union activity and investigation of company •
Missing evidence and unsolved riddle.**

In America, at least, the best-known tragedy of an individual employee in the nuclear industry must be that of Karen Silkwood. Her death in a car crash one night in 1974 could have been passed over and ignored as just one more sad traffic accident. But it was not. Hundreds of articles and a dozen books and documentaries told the story of her life and death. Finally a major motion picture, *Silkwood*, was made. Karen

Silkwood became a martyr for anti-nuclear activists, feminists and labour unionists.

Karen Silkwood was a slim, dark-haired and attractive Texan who, at the age of twenty-eight, worked as a technician at the Kerr-McGee Cimarron plutonium plant in the town of Crescent, Oklahoma. The plutonium was manufactured at the Hanford weapons production plant in Washington State, then delivered to Cimarron in heavily guarded armoured vans. The Kerr-McGee workers in Cimarron processed and converted the plutonium into pellets which were welded into fuel rods. These were then sent in heavily guarded armoured vehicles back to Hanford where they were to be used in a new experimental Westinghouse fast-breeder reactor.

In retrospect, many found the armoured security of the dispatches to and from Kerr-McGee's Cimarron plant somewhat ironic. Each vanload contained a relatively small amount of plutonium, yet the plant, which was licensed to hold up to 700 pounds of plutonium at any one time – the equivalent of fifty atomic bombs – had no comparable security. A considerable amount of uranium and plutonium simply went missing. In one notable case, a worker gave his child a uranium pellet to take to school so that everybody might see what kind of work was done at the big factory in Crescent.

Plutonium was relatively new to most Americans at that time. Few workers seem to have been told, or knew, that there might be a connection between contamination by plutonium and cancer, and the lack of concern in its handling was quite remarkable.

A couple of armed guards were finally hired in 1974 to cover the front gate, as a security measure, but it appears that they were not effective at monitoring or checking for theft. It later emerged that no security check was run on the armed guards themselves, and the very first guard hired was later revealed to be a convicted armed bank robber.

During Karen Silkwood's time at Cimarron, Kerr-McGee had a multi-million-dollar contract to produce some 16,000 plutonium rods for fast-breeder reactors. Karen Silkwood's job involved inspecting the plutonium pellets, and cleaning and polishing the welds in the rods in order to check for cracks and faults.

The Cimarron plant had a weak union presence, but Karen Silkwood's rising concern for what she saw as poor safety practices, and even poorer training of staff, resulted in her serving on a union committee where she became the safety officer. Ironically, this appointment was followed by numerous incidents in which Silkwood herself became contaminated with plutonium.

The contamination of workers, necessitating highly abrasive shower scrub-downs, was fairly common in Cimarron, but the sudden frequency of Silkwood's contamination was unusual. Finally, an extreme case of contamination occurred, not at work but in her own home. It seems that someone had spiked the food in her refrigerator with plutonium. Not knowing this, Silkwood ate the food and the ingested plutonium was not detected until she went to work the next day. Although the company maintained that Silkwood had contaminated herself just to embarrass them, this seemed hardly credible. In fact, Silkwood was terrified that she would die from plutonium-induced cancer.

At roughly the same time as the contaminations, Karen Silkwood claimed to have found documents that would prove Kerr-McGee was not being thorough in its quality control inspections and was allowing defective plutonium fuel rods to be sent out for use. This was a serious allegation, for such faults might lead to an accident in a nuclear reactor. Asked by a Washington union official if she was willing to go public with the information, she said she was.

The thirteenth of November 1974 – a week after she had eaten the plutonium-spiked food – was the date set for her meeting with the union official and a reporter from the *New York Times*. Silkwood was to have met them at a nearby Holiday Inn shortly after eight in the evening. She left an earlier meeting at the Hub Café in Crescent soon after seven o'clock in order to keep her appointment at the Holiday Inn. She never arrived.

Silkwood's car was found at around 7.30 p.m. seven miles south of Crescent having swerved off the road and smashed into a concrete culvert wall. She was crushed inside the car and must have died instantly. The union official and the *New York Times* reporter waited for her at the Holiday Inn until late in the evening, when they heard of her death.

When they finally went to inspect the towed away wreck of the car later that night, they could not find any of the papers and photographs Karen had promised to bring with her. Yet, at the Hub Café hours earlier, work mates had noticed that she had with her a reddish-brown spiral notebook and a large dark-brown folder filled with papers and what looked like photographs. She had even mentioned these documents to a co-worker minutes before getting into her car, explicitly stating that they were 'all the proof concerning falsification of records', and that she was on her way to show them to someone from the *New York Times*.

Karen Silkwood had definitely taken the folder and notebook with her in the car, but they were not found in the wreckage, nor at the scene of the crash. The union official later found out that Kerr-McGee employees, claiming the need to check the vehicle for radiation, had inspected the wreck before he had reached it.

The cause of the crash was never resolved. Karen Silkwood was an excellent driver, she had won prizes for car-racing. This was a road she had travelled nearly every day for a couple of years. Initial reports stated that she was under the influence of drink and drugs and had simply fallen asleep at the wheel. But the results of the autopsy proved the intoxication charge false and, in any case, it was later demonstrated that a car with a sleeping driver could not have drifted off the road and ended up in the position it did. It is also unlikely that, after being quite animated at the Hub Café, Karen would have fallen asleep at the wheel only seven miles down the road.

The conspiracy-theorists believe that Karen Silkwood was murdered by being forced off the road because someone knew about those incriminating papers. They believe that the documents were taken from the wreck either then or later that evening. The condition of the vehicle gives this theory some credibility, for the car sustained unexplained dents on the left rear side. These dents were not on it before the crash and could not have been caused by a single-car collision or by the tow truck that took the wreck away. They are, however, compatible with the theory that a vehicle rammed Silkwood from behind and knocked her out of control. However, the investigation into the affair failed to resolve the mystery.

3

Murder of a Rose Grower

The murder of Hilda Murrell, 1984 • English rose-grower and pillar of society • Sizewell nuclear inquiry • Falkland war-nuclear submarine connection • Commander Green's 'State crime'.

On 24 March 1984 the body of a woman was discovered in a copse near Shrewsbury. It was a well-known and respected 78-year-old woman named Hilda Murrell. Though far less widely publicized than the case of Karen Silkwood a decade earlier, there are strong parallels between the two tragedies. However, the English case proved to be even more Byzantine and far-reaching in its implications. Furthermore, if there was an element of doubt as to whether Karen Silkwood was intentionally killed or not, there was none at all in the case of Miss Murrell. She was certainly murdered, and murdered in a very brutal manner.

Hilda Murrell was one of that special breed of upright, strong-willed and independent Englishwomen, usually the 'maiden aunt' of a good family, with a quality education and an intrepid spirit. She was the kind of person one might expect to turn into an elderly detective like Miss Marple or some other comparable figure from popular fiction.

Hilda Murrell ran a family rose-growing business for many years. However, not content merely to run a successful business, she became a champion grower, winning awards at national levels in all the major flower shows and becoming a world authority on miniature, old-fashioned, and shrub roses. In fact, she bore a modest claim to celebrity status in rose-growing circles, and one breed of pink rose bears her name.

Hilda Murrell was a pillar of society. In her youth she attended Cambridge University and graduated with a Master's Degree. During the war she worked for the Jewish Refugee Children's Society. She was a founder of a local nature conservation trust and, in Wales, where she had a cottage, was on the committee of a nature reserve.

Perhaps it was a natural progression from her concern with horticulture and nature conservation that, in her later years, brought Hilda Murrell into the debate about the presence of nuclear power stations and nuclear weapons installations in England's 'green and pleasant land'. Whatever her reasons, and despite a naturally conservative temperament, Hilda Murrell became deeply concerned over what she saw as dishonest government statements and public lies on all matters relating to the nuclear industry. When the issue of the construction of the Sizewell B reactor arose in the early 1980s, Hilda Murrell decided to prepare what she called an 'Ordinary Citizen's View' on the matter.

Hilda Murrell was a very exceptional ordinary citizen, however, and after reading the government White Paper on nuclear waste, she decided to concentrate her investigations in that area. She went about it in a typically thorough way. She read massively on the subject, attended university classes on nuclear physics, and made a number of contacts with government and academic experts in nuclear power. Through the winter of 1983–4, Hilda Murrell worked continuously on her paper. Three days before her death she told friends that she had finally finished the Sizewell paper.

Strangely enough, nearly four weeks before her disappearance, Hilda Murrell seems to have been extremely concerned about her personal safety. At the end of February she made a panic telephone call to Gerard Morgan-Grenville of a Welsh-based ecological research organization called EcoRopa whom she had come to know through her research work. Although she refused to discuss the reason for her alarm, Morgan-Grenville remembered the call vividly because he found it very out of character for a woman who had previously struck him as both down-to-earth and extremely reserved. In that conversation, Murrell said: 'If they don't get me first I want the world to know that one old woman has seen through their lies.'

Less than ten days before her death, Murrell told at least two other people about her anxieties.

In a country where very few violent deaths remain unsolved by the police, Hilda Murrell's murder is full of unexplained circumstances. The police investigators maintain that she surprised a burglar who robbed and murdered her. Initial reports stated that the house had been ransacked, she had suffered multiple stab wounds and had also been sexually assaulted.

In fact, the murderer had not ransacked the house, and there was no convincing evidence of sexual assault. The stab wounds to the abdomen were not deep and evidently did not cause her death. The coroner concluded that Hilda Murrell had died of hypothermia, although the report was not released to the public or to her relatives.

There were a number of points that made the burglary theory unlikely. There was evidence of a struggle in two rooms, but the intruder seems to have carefully searched the house without notable disturbance, and without taking any objects of value. It does appear, however, that Hilda Murrell's notes and files on her nuclear investigations were carefully sorted through, and it was later discovered by a nephew that although working versions of her Sizewell report remained, the final draft had vanished.

The telephone in Hilda Murrell's home had been disconnected at the junction-box beforehand in a very sophisticated way. The result was that the line was dead for any outgoing calls, but anyone calling in would simply hear ringing tones and think the phone was in order but the owner not at home.

The greatest mystery of the case, however, was why the interloper had risked detection by removing the victim from the house. Surely an ordinary burglar would have simply fled the scene and left the body where it was?

Although police maintain that Hilda Murrell was probably knocked unconscious or killed shortly after she was last seen at noon on 21 March, her body was not discovered until 10.30 on the morning of 24 March in a copse six miles away on the other side of Shrewsbury.

Why was Hilda Murrell, dead or alive, taken from her house? Why would anyone risk being seen taking her, or her body, out to the Murrell car and driving through the busy town centre of Shrewsbury? (In fact, over sixty people came forward to say they had seen the car that day.) Why would the murderer bypass a wooded area described as ideal for hiding a body in, and drive on to an open farming area where the car was abandoned? There, police maintain, the murderer took the chance of no one seeing him carry or drag Hilda Murrell over half a mile of open ground before reaching the copse in a field where she was eventually found.

Hilda Murrell's working copies of her report were edited by others, and some time after her death it was distributed and published. It is an intelligent and well-ordered attack on the government's White Paper on Radioactive Waste Management. In her report, Murrell pointed out: 'There is now a quarter of a ton of plutonium in the Irish Sea. It would be beyond belief had it not happened, that it could ever have been thought permissible to throw this stuff into the living environment without any proper scientific examination of the results of such action.' And later: 'There has been either a total failure to monitor the place properly, or else eyes have been deliberately closed so that costs of waste treatment should not become an embarrassment to the industry.'

Miss Murrell's paper is a useful source of information but it does not appear to contain any facts that an intelligent, thorough, and inquiring citizen could not have gathered. Murrell does not seem to have had access to any secret information or leaked documents. Certainly, such organizations as Greenpeace have provided the press with far more damaging documentation in the nuclear field.

There seems no doubt that Hilda Murrell was under surveillance in one form or another, but the reason for this is a mystery in itself. It later emerged that a large number of Sizewell objectors were under similar surveillance by private investigators as well as government agencies but, so far as Hilda Murrell was concerned, it does not seem likely that any serious action could have been ordered against her on the basis of what her report contained.

However, Sizewell was not the only point of connection with Miss Murrell in which security people might have been interested. There was a military aspect to her involvement with the whole nuclear issue that Hilda Murrell herself was probably largely unaware of up to the time of her death. To many, it now appears that she was not killed because of Sizewell, but because she was believed to have information of considerable secrecy about the nuclear submarine *Conqueror*, and its role in the sinking of the Argentinian ship, *Belgrano*.

The fatal link was Hilda Murrell's closest relative, her nephew, Rob Green. The year before Hilda Murrell's death, he had been Commander Robert Green, a naval Intelligence officer during the Falklands War.

After his aunt's death, Commander Green retired from military service and began his own private investigation into all aspects of the murder. Green is a level-headed military man, not prone to conspiracy theories. However, he does possess a considerable knowledge of the workings of the British Intelligence community.

After two years of looking into the matter, Green concluded: 'I believe it was a State crime. I sense that Hilda, despite being a law-abiding citizen, became a casualty of an expansion of the activities of the State's internal security apparatus.'

Green and others point out that Murrell was killed at a time when the panic over leaks concerning the nuclear submarine activities were a major embarrassment to the Thatcher government. Questions asked in Parliament by the Labour MP Tam Dalyell had made it obvious that he was in possession of information that was classified as secret. It appeared at the time that the source of that information might have been naval headquarters at Northwood, which received and decoded all radio activity during the Falklands conflict. Commander Green was in a key Intelligence post at Northwood during the conflict. Furthermore, Green was one of only two officers there who chose to retire in the months that followed the Falklands War.

It seemed obvious to Green that in any investigation into leaks he would certainly be checked out as a possible source, more especially as he was often in contact with a close relative who was known to be an anti-nuclear campaigner. In fact, Green was not the source of the leaks, and it was later revealed that a large number of Northwood personnel were investigated as a result of the leaked material.

It does not seem entirely coincidental that, after Hilda Murrell's death, the St Albans flat of Peter Hurst was broken into. Hurst was a colleague of Green's (and the other officer to resign from Northwood). In this 'burglary' nothing was stolen but Hurst's papers were searched.

Tam Dalyell himself maintains that Murrell was kidnapped and killed in a bungled operation by British Intelligence investigating the leak over the *Belgrano* incident in which almost four hundred Argentinians lost their lives. Because of the government panic over leaks, he believes investigators were given a long rein. It seems likely that if Green was thought to have secret papers, he might well have chosen to hide them in his aunt's home.

Commander Rob Green also believes it was this supposition which led to the investigation of his aunt, rather than the Sizewell material, although her involvement in anti-nuclear activity probably contributed to the security service's view that both he and his aunt were enemies of the State.

Several aspects of the case, already mentioned, support Green's theory. He also came up with some new evidence. A neighbour of Hilda Murrell, Laurens Otter, disclosed that she had talked to him only hours before she was abducted. She had told him that a police inspector (who seems to have given a false name) was about to arrive from London to question her. No policeman has come forward on this matter; nor have the police, who were seen by neighbours inside Hilda Murrell's home on the evening of the 23rd. Local police swear they did not enter her home until the 24th, the day her body was found.

As to the timing of the killing, a local farmer inspected the copse twenty-four hours after Murrell's car had been abandoned nearby. He was marking trees for felling, and claims there was no body in the copse at that time. He was convinced that, at that late time of the year, there was no possibility of the body of a rabbit escaping his attention, let alone that of a woman.

The car itself had been examined by local people and the police a couple of hours after it was abandoned on 22 April. It contained nothing. So where was the body?

At four o'clock in the evening of the following day, a dark-coloured car was observed near the copse. A man was seen walking over to the copse, then back to the car. Later that night, lights were seen moving about in the copse.

Green believes that his aunt did not die until the day after her abduction. He thinks she was taken off for interrogation and subsequently murdered, her body being dumped by the man or men in the dark car in the late afternoon of the 23rd.

Green's theories tie in well with another private investigation launched by Hilda Murrell's friend, Dr Don Arnott. Dr Arnott is an eminent pathologist. He is also a former consultant to the International Atomic Energy Agency and an authority on nuclear energy in medicine.

There were several points which Dr Arnott found unsatisfactory in the official version of events. As a pathologist, he was not at all convinced by the results of the post-mortem. First of all, a ten-inch kitchen knife found near the body, which police identified as the murder weapon, could not, according to Dr Arnott, have caused the stab wounds sustained by the victim. This was later confirmed by the doctor who conducted the post-mortem.

Secondly, the victim's stab wounds were superficial, the worst being less than half an inch deep. These wounds were not serious enough to kill, nor even to cripple the very fit Miss Murrell long enough for her to have died of exposure in the copse, as was suggested.

Thirdly, Dr Arnott expressed concern that no toxicity test was conducted on the body. Since the external causes of death were so questionable, such a test should have been conducted to rule out other possibilities of death – such as poisoning.

After investigating the matter Dr Arnott constructed a likely sequence of events that adds further substance to Commander Green's theories. Dr Arnott does not accept that the burglary and sexual assault elements are in the least convincing. He believes there were several other red-herrings involved, like the false clue of the kitchen knife. He suggests that the Murrell car was purposely driven as a decoy through the town and beyond, while a second vehicle (a green Range-Rover seen by neighbours) took Murrell in another direction. He thinks it possible that Murrell was either drugged or poisoned by an injection in the arm or abdomen, and that the knife wounds were inflicted in order to hide the hypodermic needle-marks. He, too, maintains that Murrell died at the hands of her interrogators elsewhere, and was later dumped in the copse near her car. Whoever he was, Hilda Murrell's killer was clearly not an ordinary burglar. Commander Green has not indicated which agents he believes might have been at work in Murrell's case, but he does see the unmistakable hand of a 'covert security organization' in the affair. Speculation is that this may have been the Special Branch or MI5. Some believe Murrell's murder resulted from an operation that went wrong through the initial bungling of one of the private investigators who are known to be hired to carry out particularly illegal jobs for government agencies. Others suggest that Britain's 'nuclear police', the super-secret Atomic Energy Authority's Special Constabulary, may have been involved in the killing.

It seems likely that Hilda Murrell's murderer will never be found. Whatever the exact sequence of events, and whoever was responsible for her death, Hilda Murrell – like Fernando Pereira in New Zealand and Karen Silkwood in Oklahoma – was an ordinary citizen who dared to find fault with the nuclear industry and paid for that criticism with her life.

4

From Greenham Common to Bataan

Greenham Common, UK, 1984 murder of Dierdre Sainsbury • Bataan, Philippine anti-nuclear group: two police killings • German and Japanese protest deaths • Basque ETA anti-nuclear assassins • Canadian anti-Cruise bombing and Pennsylvania missile sabotage.

One of the most determined anti-nuclear protests in the world has been that of the women of Greenham Common, who have for years encamped themselves around the US nuclear missile base in Berkshire, which was the first to deploy cruise missiles. This brave peace camp of women activists has kept the nuclear arms issue continually before the British public eye. Among its many dramatic actions was the call in 1982 for 30,000 women to form a human chain around the base, and another similar protest in 1982 when a human chain of at least 75,000 demonstrators stretched from Greenham to two other nuclear weapons centres in Berkshire.

There has been, however, continuous harassment of the peace camp by government and local officials, as well as by soldiers and local residents. Numerous assaults, night raids and arson attacks have taken place. On one occasion, a vehicle was driven through tents occupied by sleeping women.

The most extreme incident occurred some nine months after the murder of Hilda Murrell, on 22 December 1984, when a young Greenham common protester, Dierdre Sainsbury, was given a lift by a local man and then beaten, raped and murdered.

Dierdre Sainsbury's murder was not an overtly political act. It was, however, committed in an atmosphere of intimidation largely created and sustained by government and judicial tolerance of continued assaults upon the peace-camp women. It was an atmosphere not unlike that of the civil rights marches and peace camps of the 1960s in the southern United States. Local hostility was aroused by the intruders and law enforcement looked the other way when violence flared up.

Some four years before the murder of Diedre Sainsbury, on the other side of the world, citizens in the Philippines were protesting against a controversial nuclear reactor. The reactor, ordered completed by President Marcos, was spectacularly located at Bataan in an earthquake fault, on the slopes of an active volcano. A group was formed under the considerably understated title 'Concerned Citizens of Bataan'. Although it was a perfectly legal organization, Marcos' police employed needless violence when seeking out its leaders and in separate incidents shot two of the 'concerned citizens' dead in the midst of interrogations.

Most commonly, however, deaths in the nuclear protest movements have resulted from violent police suppression of marches and protest rallies. In 1982, for instance, German and Japanese nuclear protesters were killed on two occasions when police vehicles attempted to force their way between marchers. In 1986, another German protester died in Bavarian demonstrations against the construction of Germany's first nuclear waste reprocessing plant.

There have also been a few extremists among nuclear opponents. The most violent (and, unfortunately, the most effective) anti-nuclear campaign was waged by the Basque separatist organization ETA, to stop the building of a nuclear power plant near Bilboa. After a three-year period of terrorist attacks, culminating in 1981 with the kidnapping and assassination of the plant's chief engineer Senor Jose-Maria Ryan, the ETA succeeded in achieving its objective.

Similarly, in October 1982, a Vancouver-based Canadian organization called Direct Action used a truck packed with dynamite to blow up the front offices of the Litton Systems plant in Toronto in protest against the fact that the plant makes guidance systems for cruise missiles.

However, most protest groups point out that ETA, for all its anti-nuclear stance, is not conspicuously committed to the pacifist and ecological issues which are in line with their own ideals. Furthermore, the tactics of extremist groups like Direct Action are all but universally rejected by them.

Even the most radical of the established anti-nuclear groups are, it seems, only willing to lend their support to actions which involve no risk to human life, such as that of the 'Plowshare Eight' in September 1980.

In this action three priests, a nun and four other members of the Atlantic Life Community entered the plant of General Electric in King of Prussia, Pennsylvania, where Mark 12-A missile warheads were being assembled. Here the group destroyed two warhead cones with hammers and poured their own blood over the weapons and tools.

For the majority of anti-nuclear protesters, however, even acts of sabotage are too extreme to condone. They choose to limit their civil disobedience to more symbolic actions such as clipping a single strand of barrier wire on nuclear installations, then dutifully handing themselves over to the authorities for arrest. They consistently reject the idea of fighting bombs with violence, however provoked they may feel. They choose instead to embrace Gandhi's ideas of change through civil disobedience and their message is the same as his: 'There is no way to peace. Peace is the way.'

5

Broken Arrows

Cambridge as Hiroshima: 1956 • Accidental bomb drops: New Mexico, North Carolina • Nuclear submarines: loss of 3 US and 4 Soviet subs; 350 die • Accidental launches, explosions, deaths • Greenland, Spain, Nevada, West Germany, Finland, Vietnam • 60,000 warheads: 3.5 tons of TNT per person.

On 27 July 1956, an American B-47 bomber flying from Nebraska crashed upon landing at RAF Lakenheath, Cambridgeshire. The American plane burst into flames as it struck a storage igloo. Blazing fuel engulfed the igloo which contained three Mark-6 nuclear bombs, each many times larger than the Hiroshima weapons.

In fighting the fire all four crew died but, even more alarmingly, the fire spread within the igloo. Firemen worked frantically, pouring massive amounts of extinguishing foam on to the flames which scorched and badly burned the nuclear bomb casings.

By current standards, the Mark-6 bombs were primitive devices, twelve feet long and six feet in diameter. They each contained eight tons of TNT as part of their trigger mechanism. It appears that the temperature of the flames came within a few degrees of exploding the TNT and triggering the bombs, before the firemen managed to extinguish the fire.

One US officer who was an eyewitness admitted that disaster was avoided only by 'a combination of tremendous heroism, good fortune and the will of God'. Another, a retired US Air Force general who was then a pilot at Lakenheath, stated that if the fire had ignited the TNT, 'it is possible that a part of eastern England would have become a desert'.

The incident at Lakenheath, in US military jargon, is known as a 'broken arrow' – a serious accident involving nuclear weapons. We know of about thirty-six Broken Arrows since 1945, none of which have been willingly reported by the nations involved. There have undoubtedly been many more times this number.

One of the most serious Broken Arrows occurred in 1950, six years before the near-disaster in Cambridgeshire, on the relatively remote Fairfield-Suisun US Air Force Base in California. The accident at the base – now called the Travis Air Force Base – occurred when an aircraft crashed with a nuclear bomb on board. As with the British incident, rescuers attempted to put out the blaze, but they were not as lucky as the men at Lakenheath. The bomb exploded and a total of nineteen airmen and rescuers – including General Travis – were killed in the blast. For thirty-one years, the Pentagon covered up the incident as best it could. It was not until 1981 that the military was forced to admit that the blast that had killed those nineteen servicemen came from a nuclear weapon.

Another incident involved the accidental dropping of a Mark-17 hydrogen bomb near Albuquerque, New Mexico, in 1957. This weapon is believed to have been identical to the Mark-17, code-named 'Bravo', which was dropped on Bikini Atoll in the Pacific in 1954. 'Bravo' was the largest hydrogen bomb explosion in history: fallout covered over seven thousand square miles and it had an explosive yield of 15 megatons – roughly twelve hundred times the size of the Hiroshima bomb.

The Albuquerque bomb fortunately fell on uninhabited land owned by the University of New Mexico, just a few miles south of Kirtland Air Force base. When the bomb hit the ground, the twenty tons of conventional explosives – which were used to trigger the nuclear device – ignited. The explosion blew a crater in the earth twenty-five feet wide and twelve feet deep but by extreme good fortune the unarmed nuclear charge failed to explode.

Just one year later, in March 1958, an almost identical incident occurred when a B-47 bomber dropped a nuclear bomb on to the small village of Mars Bluff, near Florence, South Carolina. Again the nuclear device did not ignite, but the explosion of the non-nuclear material demolished the house of a fortunately absent farmer and blew an impressive hole some seventy-five feet wide and thirty-five feet deep in the ground. It was an incident which left a considerable impression on the people of Mars Bluff.

Twenty years after this event, a third accidental dropping of a nuclear bomb was acknowledged by the Pentagon. In 1961, a B-52 bomber jettisoned an even more massive, 24-megaton bomb over Goldsboro, North Carolina. Evidently, five of the bomb's six safety devices were knocked out when it hit the ground. Only one remaining device prevented the igniting of a bomb nearly two thousand times more powerful than that of Hiroshima.

Two of the worst peace-time disasters in US naval history occurred on nuclear-powered and -armed submarines during the 1960s. On 10 April 1963, the USS *Thresher* was lost two hundred miles off the New England coast, while on 21 May 1968, the smaller attack submarine, USS *Scorpion*, was lost off the Azores. Inadequate concern for safety measures and procedures was the judgment levelled against the US Navy for the loss of 228 men in these disasters at sea.

Statistics on the Soviet military are, of course, extremely difficult to obtain. However, we do know that there have been a number of nuclear submarine accidents, usually fires within reactor engines which have resulted in the loss of the ships, their reactors, their nuclear armaments, and at least some crew members.

In 1961, a fire aboard a nuclear-powered and -armed Soviet submarine in the Baltic Sea resulted in the death by radiation poisoning of an undisclosed number of crew members. In 1980, a Soviet Echo-1 class submarine caught fire and sank off the coast of Japan. At least nine crew members are known to have died in the blaze. The worst known peace-time Soviet submarine disaster occurred in June 1983, when a nuclear submarine sank in the North Pacific. All 90 crew members are believed to have perished.

The most recent reported fatal accident aboard a submarine occurred on 3 October 1986, when a Yankee-class nuclear submarine patrolling off Bermuda caught fire and an explosive blast blew a hole in the vessel. At least three crew members died in the blast, and the vessel itself, armed with sixteen SS-N-6 ballistic missiles, sank soon after the rest of the crew were rescued by a Soviet merchant ship.

Among the serious accidents with nuclear weapons which have resulted in considerable radioactive contamination, America was forced to acknowledge two by virtue of the fact that they were too widely witnessed to be covered up. One occurred near Palomares, Spain, on 17 January 1966. It involved a B-52 bomber carrying four nuclear bombs which collided with a jet tanker. The conventional explosives in two of the hydrogen bombs ignited and blew the fragmented plutonium cores over a large area. Because of the plutonium contamination, a massive clean-up operation was attempted. Nearly two thousand tons of topsoil was removed and put in steel containers which were then sent off to America for disposal.

The other 'admitted' accident also involved a B-52 carrying four hydrogen bombs. This bomber crashed on 21 January 1968 near Thule in Greenland. The blast of the bombs' conventional explosives blew the smashed plutonium cores of all four bombs over an even greater area than the Palomares incident. The deadly plutonium debris was reportedly cleaned up in the most scrupulous manner. Official reports claimed that a quarter of a million cubic feet of debris and contaminated material was removed to US storage sites. Nearly twenty years later, however, Greenland Eskimo leaders have shown that these claims were largely fictitious – a military propaganda exercise to dupe the outside world into believing that such contamination could be easily cleaned up. It appears that the effects of that plutonium contamination are only now beginning to be widely recognized in local communities.

A more recent and possibly even more disturbing Broken Arrow incident occurred in 1980 when a Titan surface-to-air intercontinental missile ignited. Time and again, the US military had claimed that the accidental launching of such missiles was an impossibility. They were confident that the numerous built-in fail-safe systems covered every possible scenario. If it were not for the lethal consequences, it would be almost a laughing matter that these fail-safe systems were by-passed when an airforce technician accidentally dropped his wrench down the silo. The wrench struck a Titan II missile fuel tank, causing a leak and an immediate explosion. One man was killed and twenty-one injured as the missile launched itself, hurling its mercifully unexploded warhead a quarter of a mile out of the silo.

Again, this was not the first accident of its type. Two earlier nuclear missile accidents have only recently been acknowledged. In December 1964, at Ellsworth Air Force Base, South Dakota, a Minuteman missile on strategic alert accidentally fired its retrorocket and caused a massive explosion. Four years earlier, at McGuire Air Force Base, New Jersey, a Bomarc missile exploded within its shelter. The blast was followed by a disastrous fire and severe plutonium contamination of the area. During the clean-up operation which followed, the launch shelter was buried beneath earth and cement. An area of half-a-million square feet was then sealed off around it with a layer of concrete.

Broken Arrows are not the only category of accident involving nuclear weapons. The American military have a classification for less serious accidents. These are called 'Bent Spears' – and there have been hundreds of them. One Bent Spear in January 1985 resulted in the death of three US soldiers in Waldheim, West Germany, when the solid fuel of a Pershing II missile ignited. The incident was categorized as a Bent Spear rather than a Broken Arrow because the Pershing II missile was not equipped with a warhead at the time.

However, the Pershing II followed hard on the heels of a real Broken Arrow. Only a few days earlier, the Soviet Union had accidentally launched a test cruise missile which went out of control and shot over the Soviet-Finnish border. The missile crashed well within Finnish territory in a fortuitously remote frozen lake.

It sometimes appears that the most likely cause of nuclear war will be a human or computer error. We know, for example, that the Americans have launched nuclear weapons by accident on three occasions. Unnervingly, two of these were in Vietnam during the war there. We also have authoritative evidence that twelve nuclear bombs or warheads have simply been lost and never recovered by the military.

The arms race costs the world over $700 billion a year. There are over 60,000 nuclear warheads worldwide which amount to an explosive firepower of 3.5 tons of TNT per person on earth. Peace in the nuclear age is a fragile thing indeed.

6

Soviet Disaster: Chernobyl

Soviet odds against meltdown: one in 10,000 years
• 25 April 1986 explosion and fire: 31 deaths •
Evacuation of 135,000 • Estimate of long-term
cancer deaths: 39,000.

Following the line taken by every nuclear state in the world, the Soviet Ukraine Minister of Power, Vitali Sklyarov, in a leading article on the Chernobyl nuclear power plant in *Soviet Life* magazine, scoffed at the idea of a meltdown as impossible – not just at Chernobyl, but throughout the Soviet Union. The minister backed up his case with reams of statistics and computerized calculations provided by scientists, engineers, and experts in the nuclear industry that proved conclusively that their technology had taken into account all possible elements of error. The minister felt totally confident in his assurances. Furthermore, he confided, there was the back-up of government emergency plans – not that they would ever be needed – which, beside providing for the population, made certain that 'the environment is also securely protected'.

By the standards of optimism professed by Western politicians, the Soviet minister's assurances were modest. 'The odds of a meltdown are one in 10,000 years,' he reiterated in a statement issued in February 1986.

Two months later, on 25 April, technicians at the Chernobyl complex began a safety test on the Number Four reactor. The technicians were trying to determine just how long turbine generators would continue to produce power as a result of inertia, if the flow of steam from the reactor was suddenly halted by an unexpected shut-down.

In order to conduct this test the technicians, over a twelve-hour period, slowly reduced the reactor's power to 7% of its capacity. Because it was necessary to run the reactor at such a low level, in order to prevent the automatic-control of the emergency cooling system from cutting in, the technicians had to disconnect the system. When power began to drop too low for the test, they brought it back up by removing all but six or eight of the reactor's control rods. A second safety system also had to be shut off because it was automatically triggered to shut the reactor down when the turbines slowed and stopped, and this would certainly have prevented the test from being carried out. Finally, just before the test was due to begin, the flow of cool water to the reactor was reduced to bring down steam production.

At exactly 1.23 a.m. on the morning of 26 April 1986 the fatal Chernobyl experiment began. The technicians cut off the power to the turbine. Once the switch was thrown, the situation was instantaneously out of control. There was no time for the technicians to make any response at all. With all emergency cooling systems cut off, there was no way to control the instant and explosively soaring temperatures. This was a runaway reactor.

It was over in seconds. There was a massive surge in power and a huge explosion blasted off the reactor's one-thousand-ton, two-foot-thick steel lid, and demolished the ten-foot-thick concrete containment walls. More than thirty fires were ignited through the complex as exploding and burning uranium fuel and control rods flared up out of the reactor like shooting stars. The reactor core and its graphite elements began to burn at temperatures nearing 3,000 degrees Fahrenheit. It was a radioactive volcano that, in the days to come, would pour out more nuclear fallout than a thousand Hiroshima bombs.

As Swiss nuclear scientist Rudolf Rometsch stated several months after the disaster, the Chernobyl reactor went from 7% capacity to 'a hundred times its nominal value in less than a second, and the power plant was destroyed in the following four seconds'.

The explosion and fire sent radioactive dust high into the upper atmosphere. Firefighters in helicopters worked under deadly conditions to extinguish the blaze. They airdropped five thousand tons of extinguishing sands, made up largely of dolomite, boron and lead, on to the reactor in order to smother its flames. Yet it burned on for twelve days before it was brought under control.

Even so, this did not mean that the reactor area was safe. Fallout of lethal ash at the Chernobyl site itself was equal to forty Hiroshimas. There were 31 initial acknowledged deaths at the site of the disaster (the immediate death toll would have been much higher had the accident not occurred in the middle of the night). Over 135,000 people from a 300-square-mile area had to be evacuated from their homes, nearly all of whom were permanently relocated in new homes in safe areas hundreds of miles away.

Beyond the initial fatalities, it is estimated that between 5,000 and 25,000 cancer-induced deaths will result from high radioactive exposure levels among the local Ukrainian population. Those who were within four miles of the complex at the time of the accident are expected to have a 50% chance of survival. Furthermore, America's Department of Energy has estimated that about 39,000 people worldwide will also die from cancers caused by Chernobyl's radioactive fallout.

After the fire was extinguished, the biggest problem was to find a means by which the burnt-out reactor – which still emitted radioactivity some 2,500 times above permitted levels – could be contained. The five thousand tons of dolomite, boron and lead which had been dumped on the reactor was certainly not sufficient. Tens of thousands of tons of concrete began to be poured on to the site to entomb the reactor, but work in radiation suits was slow and stressful. The high levels of deadly contamination allowed workers to be active for only very short periods. Even bulldozer-, crane-, and remote-control-device operators who were enclosed in machinery with lead shielding could only work short shifts.

It then became necessary to vent the still smouldering core of the reactor to prevent this formidable concrete tomb from exploding, volcano-like, because of a build-up of heat. Work in the reactor-tomb was further slowed down because the Soviets literally ran out of cement supplies.

The formidable job of scraping away thousands of acres of contaminated topsoil, trees and vegetation around the site was yet to be tackled, and the problem of what to do with the millions of tons of contaminated material. Most critical of all was the need to create a concrete barrier to contain contaminated waterways and reservoirs in the area. Before all else, it was necessary to protect the Dnieper River which was the major source of water for the 2.5 million people of Kiev, some eighty miles south of Chernobyl.

The fire-fighting, medical services, evacuation and clean-up at Chernobyl have cost the Soviets in excess of $5 billion; and this is just a short-term financial outlay.

Many nations were indignant at the Soviets' initial reluctance to admit the extent of the Chernobyl disaster – which they did not do until nearly three days after the accident. When an account of the entire episode was presented in an Official Report to the International

Atomic Energy Agency four months later, those who read it were astonished by its tone and content.

The 430-page report, issued by the IAEA, was remarkably informative, open and self-critical. Although a few questioned the accuracy of certain points of technical detail, one anti-nuclear agency expressed surprise that the Soviets had volunteered more information about Chernobyl than they were able to discover about the nuclear accident at Three Mile Island, Pennsylvania, four months after it had occurred.

The admission of gross error in such matters was almost unheard of in Soviet reports to the outside world. Yet here was Andronik M. Petrosyants of the Soviet Atomic Energy Commission stating: 'The accident took place as a result of gross violations of operating regulations by workers.'

The Soviet report cited six major errors in the Chernobyl incident:
1. The emergency cooling system was disconnected.
2. The reactor power was accidentally lowered too much, making it difficult to keep under control.
3. All but 6–8 control rods were pulled out of the core – the absolute minimum should have been 30.
4. All water pumps were on and exceeding permitted flow rates.
5. The safety device that shuts down the reactor when water levels change rapidly was cut out.
6. Most critical of all, the automatic-control safety system that shuts down the reactor if the turbines stop was closed off.

The Soviet commission admitted that the experiment should never have been attempted, and that it was carried out by technicians with an astounding lack of understanding of how reactors worked. Chernobyl's director, Viktor Bryukhanov, and its two chief engineers, Nikolai Fomin and Anatol Dyatlov, were arrested for their part in the disaster. They were all convicted the following year and jailed for criminal negligence which resulted in manslaughter.

The incident at Chernobyl on 26 April 1986 must certainly go down as the worst disaster in the thirty-two-year history of the civil nuclear power industry. The Ukraine Minister of Power's confident prediction of one hundred centuries of safety just two months earlier was obviously in need of a little fine-tuning in its calculation. His margin of error was a factor of sixty thousand or approximately six million percentage points.

As terrible as it was, the Chernobyl disaster was not the ultimate nuclear reactor nightmare. It was technically a burn-out rather than a proper meltdown. Chernobyl was not the type of reactor that could produce the famous China Syndrome, which can only occur if a reactor contains water rather than graphite. Meltdowns occur when overheated reactors boil away the massive water reservoirs, and the molten core theoretically burns its way, mole-like, deep into the earth's crust with repercussions on a scale never experienced before.

7

The Reactors

Windscale/Sellafield accidents 1957, '76,'81,'85: 33 cancer deaths • Chernobyl 'panic' in Europe and America • Sweden's shut-down • World's 380 reactors • US accidents 1961, '68, '72, '75, '79, '86 • Nuclear dumps: 1957 USSR – hundreds of deaths • Collapse of the nuke industry in US.

Even as the Chernobyl reactor fire still blazed away, spewing massive radioactive debris on the winds that would spread it right around the planet, Margaret Thatcher saw that her first priority was to stick up for the nuclear industry. Even before she could learn exactly what had happened, she declared: 'A Chernobyl accident could not happen here.' British nuclear technology was far too sophisticated and safety-conscious. 'The record of our own nuclear power industry,' continued Thatcher without a moment's hesitation, 'is absolutely superb.'

Mrs Thatcher's statement left critics of the industry in Britain speechless. Britain's Windscale plutonium plant – later renamed Sellafield – was notorious for possessing the Western world's worst record for accidents and poor safety regulation. Windscale/Sellafield has itself now publicly acknowledged major accidents in 1957, 1976, 1981, and 1985. Even the government's own ministers have called it poorly managed and sloppily run. One of the world's largest nuclear reprocessing plants, it is also the 'dirtiest' – according to one report, it is a thousand times as 'dirty' as its nearest rival, Cap de la Hague in

France. Its dumping of plutonium has made the Irish Sea the most radioactive on the planet.

Furthermore, the 1957 fire and near meltdown of Windscale's Pile Number One plutonium plant was ranked the world's worst nuclear power station accident – before the Chernobyl disaster managed to displace it in notoriety nearly thirty years later. The fallout from the fire contaminated two hundred square miles of countryside. Milk and dairy products in these areas were dumped for months after the incident. Even the government's own conservative estimates allow that at least thirty-three deaths by cancer could be traced to radiation from the Windscale fire. Thatcher's reaction to the Chernobyl disaster was echoed by President Reagan. He too wanted at all costs to avoid attacking the nuclear industry itself. As one Reagan aide said at the time: 'We don't want the hysteria building around the Soviet accident transferring over to the American power industry.' Still less did the Americans want it to reflect on the issue of the nuclear industry's military factor.

Reaction to Chernobyl has been most marked, predictably, in countries closest to the disaster site: unfortunately, these included the Scandinavian nations who are already the most environmentally attuned in Europe. In Denmark and Norway, which have no nuclear industry, Chernobyl stirred up strong anti-nuclear sentiment which was directed at nuclear-armed Nato ships and weapons. Finland and Sweden were among the countries worst affected by the disaster, and among the most vocal in condemning the Soviets for not notifying their neighbours immediately after the explosion.

For reasons of political expediency, Poland, Romania, East Germany and other Soviet bloc countries were not highly critical of the situation on an official level. Unofficially there was a great deal of anger. Severe problems and hardships were imposed on them by contamination of crops and the banning of dairy products. Children under the age of sixteen were given an iodine solution to keep them from absorbing radioactive isotopes.

The West Germans and Austrians, however, condemned the Soviets on both official and popular levels. One monitoring body determined that West Germans may have received a radiation dose equivalent to between twenty and forty chest X-rays from the incident. Destruction of contaminated agricultural products alone cost West Germany a quarter of a billion dollars. Furthermore, anti-nuclear forces and the German Green Party used Chernobyl as a major argument in their protest against the building of the controversial Wackersdorf power station in Bavaria. The Wackersdorf station is to be Germany's first nuclear fuel reprocessing station – like Sellafield in Britain and Cap de la Hague in France. Such stations are by their very nature far more highly polluting than other nuclear installations.

Sweden's political reaction to Chernobyl was undoubtedly the most extreme of any nation. A referendum in 1980 had already determined that the public was committing the government to close down all its nuclear power plants by 2010, although, because of the cost, many leading politicians were in opposition to this measure, or at most, luke-warm in their enthusiasm to shut them down.

The political climate in Sweden was notably shaken by the threat of long-term damage in the form of increased cancer rates. A more immediate consequence of the disaster was that tens of thousands of sheep, lambs and other farm animals were contaminated because of fallout. Berries, mushrooms and many agricultural products were also unfit for consumption. And over 15,000 nomadic Lapps had their livelihood threatened through the contamination of their reindeer herds. The dumping of contaminated foods alone after Chernobyl had an initial cost to Sweden of $150 million.

At the time of Chernobyl there were an estimated 380 commercial nuclear power stations operating in 26 nations: 101 in USA, 51 in USSR, 44 in France, 37 in UK, 32 in Japan, 20 in Germany, 15 in Canada, 12 in Sweden, 8 in Belgium and Spain, 6 in India and Taiwan, 5 in East Germany, Czechoslovakia and Switzerland, 4 in Bulgaria, South Korea, and Finland, 3 in Italy, 2 in Argentina, South Africa, and Hungary, 1 in Brazil, Pakistan, Yugoslavia and the Netherlands.

In absolute numbers, Sweden with twelve stations is the world's eighth largest user of nuclear power. This country's commitment to closing its power stations is all the more remarkable when it is considered that nuclear energy provides Sweden with 50% of its total power. Compare this with the US at 17%, USSR at 11% and Britain at 18%. Exceeded only by France and Belgium, Sweden is the third most nuclear dependent nation in the world.

Reagan's concern at what he called 'anti-nuclear hysteria' over Chernobyl was, in one respect, justified. The civil use of nuclear energy for power generation in America has been in severe trouble for a number of years, although this does not seem to have extended to the dangerous military aspect of the nuclear industry.

When the Chernobyl reactor exploded, it had been only thirty-two years since the first commercial nuclear power station had come on line, yet despite assurances of the 'one-accident-in-ten-thousand-years' kind by most major American politicians and their advisory panels of nuclear scientists, there have been a long list of accidents at nuclear power plants. The first of these occurred on America's doorstep, in an experimental nuclear reactor at Chalk River in Canada, during the initial year of commercial nuclear power, 1954.

Other serious accidents at nuclear power plants have included:

1961, Idaho Falls: A military SL-1 reactor erupted and a radioactive steam explosion killed three technicians. The force of the blast resulted in one technician being impaled on the reactor's elevated control rods on the ceiling, where he remained for six days before others were able to take him down.

1968, Detroit: A fast-breeder reactor overheated. Part of the core began to melt, allowing gases to escape before it could be brought under control.

1972, New York: An explosion in a plutonium works plant did not cause fatalities, but contamination was so great that the plant had to be permanently closed.

1975, Browns Ferry: A workman using a lighted candle to check for air leaks started a fire in a cable channel that managed to knock out all five emergency fail-safe systems and nearly brought about the destruction of two twin reactors.

1979, Three-Mile Island, Harrisburg: An operator error and stuck gauge valve led to a near meltdown and an escape of radioactive gases and debris. This resulted in the destruction of the billion-dollar reactor; another billion dollars was needed to entomb the reactor and clean up afterwards.

1986, Gore, Oklahoma: Radioactive gas leak and explosion at a Kerr-McGee uranium plant resulted in the death of one worker and the hospitalization of 100.

In addition to these accidents, which are classified as high-radiation-level leakages, there have been scores of massive radioactive leakages at military nuclear plants, uranium mining facilities and radioactive waste dump sites. Such leakages are so extensive that one government expert estimated that the cost of cleaning up would exceed $100 billion – and therefore could not be considered.

The disposal of radioactive waste is, of course, an extremely contentious issue. No one has yet decided what a 'safe' disposal area for high-level plutonium, with a half-life of over four million years, would be. And as for choosing a proper site for it, the cry from every regional politician's mouth has been: 'Not in my back yard', sometimes called the NIMBY factor.

Disposal of lower-level waste has also proved a major problem. In 1969, in Colorado, a nuclear dump site spontaneously ignited, and the fire released plutonium dust over a wide area.

The worst radiation dump site accident, however, seems to have been largely hidden from public scrutiny. This occurred in the Soviet Union, and although it is difficult to reconstruct entirely, it seems likely that it was a disaster on the scale of Chernobyl.

It is now believed that a massive explosion rocked the region near Kyshtym in the Southern Urals late in 1957. The explosion was the eruption of a Soviet radiation dump. Soviet dissident scientist Lev Tumerman – who later emigrated to Israel – visited the area around Kyshtym in 1960, some three years after the incident, and claimed it was a desert. 'As far as the eye could see, there was an empty land. The land was dead: no villages, no towns, only chimneys and destroyed homes, no cultivated fields or pastures, no herds, no people – nothing', As he drove through this wasteland, he read road-signs which advised motorists not to stop for the next forty kilometres and to travel at speed.

In 1976, the exiled Soviet scientist, Zhores Medyedev, wrote an article which claimed that carelessly dumped radioactive waste had been building up on the site for years before the eruption. Medyedev believes that hundreds if not thousands may have died as a result of the disaster. Whatever the fatality rate, the devastated area was fenced off and enclosed with a canal system – presumably to stop leakage of contaminated water. By the end of the seventies, it became apparent that about thirty small towns in the region had simply disappeared from all Soviet maps.

American and British defenders of the nuclear industry immediately after Chernobyl, arguing that such an accident could not possibly happen in the West because of superior technology, went on record stating that their own reactors – unlike Chernobyl – are encased in massive steel-reinforced containment buildings. They also pointed out that neither country used the older technology of graphite-moderated reactors, and that US reactors are substantially safer because they are gas-cooled rather than water-cooled.

However, these arguments do not bear close scrutiny. Most of Britain's Magnox-type reactors have no secondary containment whatsoever. In America, one civil reactor in Platteville, Colorado, is both graphite-moderated and has no concrete encasing dome. Furthermore, none of the five weapons-production plants run by the government have concrete domes, and the oldest of these, at Hanford, Washington State, is both graphite-moderated and water-cooled.

If one were looking for a likely American nuclear disaster site, Hanford is probably one of the best choices. A leaked report made by an auditor for the facility, in June 1985, twice recommended closure of the complex's two plutonium factories because of poor design, sloppy and unsafe work practices and incomplete and false documentation. The audit discovered that essential joint welding was undersized and done by unqualified workers, inspection of welds was faulty and in many cases inspection documents were found to have been dated before the work was carried out. Hanford also has a billion-dollar problem with seepage of radioactive wastes. Despite the audit report, however, the plant remains in production today.

The containment argument was finally hit on the head when it emerged that Chernobyl did, after all, have a containment structure, and a very substantial one at that. However, it is questionable whether any practical man-made structure – in East or West – could contain an explosion capable of blowing off a thousand-ton steel lid and demolishing ten-foot-thick reinforced concrete walls.

Most observers of nuclear developments in the United States have concluded that, although Chernobyl may have put another nail in the coffin of the civil nuclear industry, the real turning-point occurred in 1979 when Three-Mile Island nuclear power station blew up in Harrisburg, Pennsylvania. Since that incident, in fact, no nuclear stations have been commissioned and the construction of several has been stopped. The most dramatic curtailment was the 1983 Washington Public Power Supply System's cancellation of four of its five proposed plants. The action by WPPSS – pronounced 'Whoops' by the wits – resulted in a default on two billion dollars' worth of bonds.

However, many have argued that, in America at least, it is not 'anti-nuclear hysteria' that has caused the demise of the civil nuclear industry. Vocal as many anti-nuke groups in America may be, the real reason for the trouble is economics. Inflation, bad management, cost overruns, safety problems have all resulted in spiralling construction and running costs. The fact is that, even in the best of times, nuclear-generated energy is astronomically expensive if all aspects of its production are taken into account. Even without the expense of clean-ups after disasters, many economists maintain that far from being the cheap source of power it is so often advertised to be, it is the most expensive.

Despite the British government's attempts to disguise the costs of nuclear power, by mid-1988 the financial reality of the industry was such that it could no longer be covered up. This realization came with the necessity of the closing down of the twin Magnox reactors at Berkeley in Gloucestershire. Berkeley is likely to be the first of eight of the older-type Magnox stations which will have to be closed over the coming years because of their age, and because they will be unable to pass safety standards.

Most of these Magnox reactors, which produce about half of Britain's nuclear electrical power, have a power-producing life of approximately twenty-five years. However, it will take at least half as much time again to dismantle them and entomb them in a concrete structure that must last for several hundred years. The cost of decommissioning a nuclear reactor is not yet known since this has never been done before, but it will certainly amount to many billions of dollars. No government has seriously addressed itself to the financial aspect of this phase of nuclear power.

The Berkeley closure comes at an embarrassing moment for Britain's Conservative government, committed to the 'big lie' that nuclear energy is critical to the nation. In the midst of attempting to privatize the electricity industry, it is faced with the reality that nobody in the private sector believes any longer that nuclear power is a viable financial investment. In the end, it will be the public, and not a private-sector company, which will pay for the government's reliance on an expensive and faulty technology.

Three-Mile Island, Chernobyl and other accidents may be factors in the collapse in public enthusiasm for nuclear power stations, but it is its own lack of cost-efficiency that has put the civil nuclear industry in real trouble.

8

Slow Murders

Low-level radiation • Nuclear weapons tests:
150,000 cancer deaths • Atomic power and
theories of 'safe' levels • Mancuso's study • Guinea
pigs: military and civilian exposure to nuclear tests
• Citizen action groups • Dr Gofman's nuclear war
crimes.

Since that first fateful 'Trinity' bomb explosion at Los Alamos on 16
July 1945, there have been approximately 1,500 nuclear tests of which
nearly 500 have taken place in the atmosphere. A UN report
conservatively estimates that 150,000 people will die prematurely from
cancer as a result of these tests. Others have sound evidence that puts
estimates in excess of 1,500,000.

The difficulty in dealing with nuclear accidents and nuclear
pollution is – in all but the most extreme cases – that the length of time
it takes for the damage to manifest itself can be years or even decades.
On the face of it, the immediate mortality rate at nuclear test sites,
nuclear power plants, uranium mines and radioactive waste dumps is
quite low. Yet, despite such exceptional disasters as Chernobyl, many
believe that the greatest threat posed by nuclear power – apart from
outright nuclear war – is in low-level nuclear pollution. The most
volatile issue in the whole debate is therefore the answer to the
question: What constitutes a 'safe' level of radiation?

Largely to calm public fears about radiation the US Atomic Energy
Commission (AEC) contracted Dr Thomas Mancuso of the University
of Pittsburgh to undertake a study of radiation on a quarter of a
million atomic workers. It was the largest and most thorough study of
its kind.

After fifteen years, Dr Mancuso published the first results of his
study. As soon as he did so he found that his funds were cut, he was
removed from the project and his data was confiscated. The results of
the study were not to the liking of the AEC – they regarded them as
disruptive to the industry. Dr Mancuso's report concluded: 'Low-level

radiation *does* cause cancer and that's definite, and the risk is anywhere from ten to twenty times greater than had been estimated. That is the essence of our findings . . . It means that lower levels of radiation, the so-called 'safe' levels, are not safe at all.'

The results of the study should not have surprised the AEC. For a large number of studies and experts had been telling them the same thing for some time – although not, it is true, with the back-up of such massive statistical evidence. One of the most unexpected Cassandras is Dr Karl Z. Morgan who became involved in the US nuclear programme in 1942 and is known, after twenty-five active years in the industry, as the 'Father of Health Physics'. Strangely enough, it was Dr Morgan who, in the 1940s, initially established the maximum permissible radiation limits allowable for nuclear workers. By the late 1950s he realized this was far too high. Today, he maintains that the 'safe' level should be 'reduced by a factor of 240'. However, the industry refuses to acknowledge the need for revisions.

Among the most vocal of the nuclear victims in recent years are those who were drafted into the nuclear tests as human guinea pigs. In large parts these were soldiers in service to their own country, who are only just coming to realize the consequences of those early 'experiments'.

By the early seventies, over three hundred thousand American servicemen had been intentionally exposed to atomic radiation in the testing of nuclear bombs. Proportionately comparable numbers were involved in similar tests carried out by most other early nuclear nations. Just what purpose was served by the exposure of soldiers in such massive numbers has never been satisfactorily explained. Thousands took no real part in the bomb experiments other than being present when they were ignited. Surely, if it were absolutely necessary to use human subjects, any meaningful results concerning human reaction to radiation could have been determined by using only a handful of men?

Even in a radiation experiment on white rats, the most elementary research student would not be permitted to experiment on a hundred thousand subjects, when equally valid results could be acquired with ten. Further, not even the most simple-minded of students would dare to declare that because the irradiated white rats did not drop dead in their tracks, the animals suffered no effects from exposure. Yet no effort was made by scientists or the military to study the long-term effects of exposure on soldiers who attended the bomb exercises. In fact, millions of dollars have been spent by the military in an attempt to prove exactly what even that hypothetical simple-minded student would not accept: the subjects did not drop dead in their tracks, therefore they suffered no effects whatever from exposure.

The truth must be faced that these hundreds of thousands of troops were not gathered at nuclear bomb test sites for the purpose of valid experimentation, any more than Thomas Edison's demonstration of the power of electricity through the public electrocution of a circus elephant was an experiment.

The troops were simply part of a massive public relations propaganda exercise to sell the bomb and the nuclear industry to the general public. Their lives were put at risk to instil pride in possession of an awesome military power that only the nuclear industry can grant a nation. The spectacle also served as a fearful reminder that only by massive and continuing financial support of the nuclear industry can that nation hope to remain in possession of superior power.

In the midst of all the controversy over the long-term effects of radiation, many have pointed a finger at the scientists of all nations who have subjected their own people to similar atomic radiation 'experiments'. Some have drawn a parallel with the use of involuntary human subjects in experiments, for which Nazi scientists were condemned at the Nuremburg war crimes trials.

One of the more remarkable individuals to draw such a parallel is Dr John W. Gofman, a radiation scientist and medical doctor who co-discovered the fissionability of Uranium 233 and who helped to refine the first milligramme of plutonium. In 1979, looking back on the nuclear issue, Dr Gofman stated as a matter of fact: 'We have already accepted the policy of experimentation on involuntary human subjects.' With painful self-criticism he continued 'I am on record in 1957 as *not* being worried about the fallout and still being optimistic about the benefits of nuclear power. There is no way I can justify my failure to help sound an alarm over these activities many years sooner than I did. I feel that at least several hundred scientists trained in the biomedical aspect of atomic energy – myself definitely included – are candidates for Nuremburg-type trials for crimes against humanity through our gross negligence and irresponsibility.' Furthermore, he concluded: 'Now that we *know* the hazard of low-dose radiation, the crime is not experimentation – it's murder.'

When it comes to the nuclear question, it is an undeniable fact that once a government is committed to nuclear power and weapons, that government is also committed to deceiving its citizens about all negative aspects of the nuclear industry. The capital investment in the industry is so massive, it seems that politicians believe there is too much at stake to do otherwise.

9

The Patriotic Cowboy

John Wayne pro-nuke patriot • Alaska tests: cowboys versus Indians • Greenpeace and Chief Dan George • *The Conqueror*: Wayne's radioactive Nevada filmset • John Wayne as nuclear fatality.

The circumstance which led to the formation of the environmental group Greenpeace in 1970 was the American nuclear bomb test on Amchitka Island in Alaska. The Vancouver-based group formed the 'Don't Make a Wave Committee' which gained massive public support in Canada in its protest against the American test. Canadians were angered that America chose to test nuclear bombs just off their coast in a known earthquake fault – especially as, just the year before, an earth tremor had triggered a devastating tidal wave in that area. Others thought that Amchitka Island was an inappropriate place to test a nuclear bomb because, quite apart from being in an earthquake fault, it was a nature reserve for endangered species.

In the midst of this protest action, that all-American film star and advocate of the bomb, John Wayne, happened to be on a fishing trip off Vancouver Island on his personal yacht, a converted US minesweeper. In an interview with the local press Wayne expressed his opinion that it was none of Canada's goddamn business what America did with its bombs. 'We need those bombs to fight the commies,' he demurred.

The world's foremost cowboy actor held exactly the opposite opinion to the world's foremost Red Indian actor, Chief Dan George, who lived near Vancouver. The chief decided to side with Greenpeace and the protesters. However, the Greenpeace activists regarded Dan George's alignment with the ecologists as a perfectly natural one. They saw the Indian and the tribal peoples of the world as protectors of the earth and opposed to the conquest mentality of the American cowboy. And certainly John Wayne was the embodiment of all that the American cowboy stood for to millions of people the world over.

It seems all the more ironic, therefore, that John Wayne has proved to be one of America's most famous victims of the nuclear bomb. In 1956, Wayne starred in what many people have listed as one of the ten worst Hollywood films of all time, *The Conqueror,* in which he played a temperamentally suitable, but highly unconvincing, Genghis Khan. Unfortunately for Wayne and other cast and crew members, the script was not the only fatal factor in this film. For the most part it was filmed on location in Nevada in 1954. The location turned out to be just 137 miles downwind of the Yucca Flats atomic test range.

By 1980, it had been determined that at least 91 of the 220 cast and crew members who worked on *The Conqueror* had contracted cancer. Out of the 91, 47 people had died from the disease by 1980. Among the dead were Susan Hayward, Dick Powell, Agnes Moorhead and John Wayne himself.

Coincidentally, in the year that followed the Amchitka nuclear protest, Chief Dan George was nominated for an Academy Award for his role in *Little Big Man.* It was a matter of some speculation that, although Dan George was heavily tipped to win an Oscar that year, he might have damaged his chances through his stance against the bomb. In fact, at this point, John Wayne surfaced and publicly stated that in his opinion – on entirely patriotic grounds – the Canadian redskin shouldn't get an Oscar.

Whatever the reason, Chief Dan George failed to win an Oscar that year. Still, while the screen cowboy met an early death through the effects of the bomb, the old rebel chief lived on comfortably and cheerfully into his nineties.

10

Contamination

Rainbow Warrior affair and paranoia of nuclear blacklists and hit lists • Nuclear industry's totalitarian élite as the enemy of democracy • Contamination – moral and physical.

The most unusual aspect of the *Rainbow Warrior* affair was that, to some degree at least, the murder mystery was solved. Although justice was not done, the truth came out. There have been many other acts of state terrorism linked with the anti-nuclear war but it is seldom possible to prove that they are linked directly to government officials.

Indeed, if the agents had not been caught red-handed, there is no doubt that Greenpeace activists would have been scoffed at for pointing the finger at the French government. Their accusation would have been dismissed as just one more lunatic fringe conspiracy theory. Why, after all, would the French government worry about the activities of a small-scale anti-nuclear protest group? Surely only a total paranoid would believe that violent action was necessary to stop such a group.

Why indeed? The French government and French military had more than enough ammunition to deal with the protesters without resorting to bombing and murder.

The answer is that when it comes to nuclear issues the French – and the governments of *all* nuclear powers – *are* paranoid. In every case where a government has committed itself to nuclear weapons or nuclear power, all those who oppose this policy are treated in some degree as enemies of the state. This has proved the case even when citizens act in accordance with their legal rights to protest or campaign against a government policy. It goes without saying that virtually all anti-nuclear groups are under police surveillance of one sort or another.

In the case of the *Rainbow Warrior*, what threat did Greenpeace present to the French military? The only conceivable one was to provide a certain amount of adverse publicity, but even this could not have been achieved without French provocation. Undoubtedly the best strategy would have been to ignore the protesters.

But the French could not bear any criticism or public censure. The bombing was an irrational and murderous act of violence but not an unusual one. Other governments and agencies have acted in a similarly paranoic way over nuclear issues, as evidenced by the cases of Karen Silkwood in the US and Hilda Murrell in Britain.

The events of the *Rainbow Warrior* and Chernobyl have not led to an improvement in attitudes, but have simply increased the paranoia of the nuclear establishment.

By late 1987, however, French journalists of virtually all political persuasions were standing up to condemn the government and the Electricité de France for refusing to supply even the most basic information on nuclear power.

Why the French government and nuclear establishment should feel so committed to secrecy on the issue of domestic nuclear power was something of a mystery to the media. There is virtually no anti-nuclear movement in France; indeed, the French population is remarkably accepting of its country's massive commitment to nuclear power.

Moreover, as some critics have stated, secrecy is not necessarily synonymous with a secure industry. In fact, as the Soviets have shown, quite the opposite is often true; an industry that is not open to public scrutiny can become uncritical of its own actions and procedures. Certainly the bizarre case of 'radioactive vengeance' perpetrated in 1981 by a nuclear worker named Noel Lecompte at Cap La Hague suggested that France's reactor sites were anything but secure. Lecompte was sentenced to nine months in prison for stealing radioactive material from the plant and attempting to irradiate his boss by placing it in the man's car. Considerable amounts of plutonium have also disappeared from a number of installations around the world.

During this widespread vocalization of outrage by a united press and media, the French nuclear industry refused to comment or reply to questions. Government officials similarly remained silent, arguing that all such matters were secret for reasons vital to national security.

Even within the scientific establishment, the nuclear industry works with a zealot's fervour to suppress the activities of scientists and academics with expertise in the nuclear field if they seem to be conducting research that might be seen as contrary to the industry's interests.

In 1987, Australian scientist Brian Martin published a partial, but extensive, nuclear dissident blacklist in *Science and Public Policy*. The list of case-studies provides evidence of a systematic suppression of informed critics of the industry. It is a depressing record of dissidents worldwide who have had their careers ruined, their funding withdrawn and their data confiscated.

The nuclear establishment acts punitively against dissident scientists because they endanger what the industry sees as its monopoly of expert opinion. If it can maintain this monopoly, which is has for so long claimed, it can continue to argue to governments that the industry alone is capable of making informed decisions on nuclear issues.

Martin and others see an urgent need to protect nuclear dissidents. Certainly, they recognize the need to support the principle of academic freedom, but, even more important, they argue that it is precisely this absence of outside criticism which results in poor management, dangerous practices, extreme imbalance in energy investment, and absolute political control over nuclear decisions.

Robert Del Tredici, in his book *At Work in the Fields of the Bomb*, interviewed the scientist William Lawless who, for five years, was the US Department of Energy's expert on nuclear waste. Soon after taking up his appointment, Lawless became aware of gross inefficiencies, engineering faults, and managerial errors that he regarded as extreme public hazards. After he had concluded his inspections of nuclear facilities, Lawless stated that the Department of Energy's establishments were so badly run that 'there isn't a Department of Energy facility in the country that could operate under commercial company regulations'.

Lawless's investigations for the Department of Energy into the military plutonium plant at Hanford, in Washington State, revealed leakages from its 149 failed high-level waste tanks. He estimated that the bill for a satisfactory clean-up of these leakages would be $12 billion and that a clean-up of all the country's nuclear facilities would currently cost America between $100–$200 billion.

Lawless believes that the autonomy and unaccountability of the nuclear élite is extremely dangerous: 'This is a closed system which is primarily responsible only unto itself, not unto Congress and not unto the American people – because they don't know enough to ask the good questions.'

At the very least, Lawless believes that the industry should be forced to accept 'small peer review groups', because 'scientists and engineers can make some very bad mistakes working by themselves'.

However, Lawless – like Brian Martin – is aware of just how hostile the nuclear establishment is to any intervention. 'The secrecy aspect of it has made many people feel that if they question what's going on, they're traitors. Somehow we've got to get people in there who really know what's going on.'

As it is now constituted, many independent observers have come to see the nuclear establishment as a totalitarian élite, as intolerant as it is powerful. Some have even come to regard the nuclear industry in Europe and America as, itself, the worst enemy of the democracies it was supposedly installed to protect.

As outrageous as some of the actions against nuclear critics have been, there is no evidence that they will not occur again in the future for the cancer-inducing qualities of atomic radiation are, it seems, both physical and moral.

12

War of the Heart and Mind

Towards a negotiated peace settlement

'We cannot command nature except by obeying her.' *Francis Bacon*

1

A Severed Head

A human head in the British Natural History
Museum • Collector's obsessions and the extinction
of the ivory-billed woodpecker, dwarf caribou, NZ
quail • Pure science as pursuit of knowledge not
wisdom • Wisdom has a heart as well as a mind.

If you are willing to search diligently through the collection shelves in
that most respected and beautiful of the world's scientific institutions,
the British Natural History Museum at South Kensington, you will
come across one of the more startling sights in that huge inventory of
planetary life – a stuffed and mounted human head.

An amazing demonstration of taxidermy as an abstract science, this
exhibit proves something of an embarrassment to the current museum
staff and is kept shut up in a cupboard. A good part of the collection at
the Natural History Museum has come to the institution from private
collections and the head, which was similarly acquired, arrived some
time during the nineteenth century.

What is disconcerting, not to say chilling, is the unknown story
behind this African man's severed head. Being a 'fine male specimen',
he seems unlikely to have died of natural causes. It seems
inconceivable today that a British amateur naturalist would travel to
an African country and, in the course of bagging numerous animals as
souvenirs of his trip, might decide to track down one of its wilder
citizens and take his head home to be mounted as a trophy for his
study. Furthermore, even if the cause of death were innocent, the
possessor of such a trophy might have considerable difficulty in
finding a taxidermist willing to take him on.

Human atrocities aside, the mania for collecting and the search for
'scientific' knowledge has led naturalists and museum collectors along
many now unacceptable paths. Like obsessed stamp collectors, some
have pursued the last of many species and extinguished them for the
sake of their collections.

A case in point is the California condor. Admirable though current
attempts by the scientific community are to save this rare bird, it was

an earlier generation of that same community that was largely responsible for the critical condition of the species in the first place. Knowing full well the rarity of the California condor, virtually every museum in the world paid collectors substantial sums to obtain numerous specimens. There were probably only 60 California condors left in the world by 1910 yet, during the previous three decades, over 200 eggs were stolen for museum collections and no less than 288 condors were actually killed for museum specimens.

I am not trying to suggest that natural historians are by nature immoral; indeed, they are currently among the most innocent in the scientific community when ethics and morality come into conflict with scientific research. The point I want to make is that the severed human head and the specimens of extinct animals demonstrate that many of our views on what is permissible in the way of scientific inquiry have radically changed during the last few centuries. However, in some circles, it still seems perfectly acceptable to 'collect' the head of a critically endangered species such as the mountain gorilla – even though there are now laws prohibiting such action.

This underlines the fact that there is no morality inherent in the pursuit of pure science. In one respect I suggest that science is itself rather like that severed head in the Natural History Museum. The human brain with all its powers of thought and perceptions is indeed a wonderful thing, but when it is cut off from the promptings of the human heart, it is dead and without purpose. It is severed from the very source of its own life.

True human wisdom acknowledges the importance of the promptings of the heart as well as of the mind.

2

The Law of the Jungle

Social Darwinism and the law of the jungle •
Miracle of jungle is self-generating life, not sudden
death • Law of jungle is harmony and diversity:
'the propeller of evolution' • Man's obsession with
dominance of nature.

Some very peculiar things have happened to the theories of Charles
Darwin since their publication over a century ago. After the mass of
the world's population eventually recovered (for the most part) from
the shock of the concept that humans are 'descended from monkeys',
many saw certain advantages in advocating aspects of evolutionary
theory. They saw in Darwin's theories a means of justifying some
particularly obnoxious social policies. These advocates of 'Social
Darwinism' usually seized on the idea that man's rise in evolution was
linked to his highly aggressive nature.

Possibly intentionally misunderstanding the concept of 'survival of
the fittest' – which in any case was not a term used by Darwin – they
perceived evolution as a justification of a 'might makes right' morality.
This – it was argued – was the 'law of the jungle' that made man the
master of the natural world. Few who advocate the theory of the law of
the jungle seem to know much about jungles, for anyone who spends
time in one is overcome, not by an awareness of violence and sudden
death, but by the great abundance and diversity of life to be found
there.

The miracle of the jungle is its self-generating life, the ability of life
to create more life. As Jacob Bronowski once stated: 'Diversity is the
propeller of evolution.' The jungle exists by virtue of the
interdependence of its life-forms, creating a harmony that allows the
widest possible diversification. It is an evolutionary laboratory that
produces more new forms of life than any other environment on the
planet.

Foolishly, men have often assumed that the profusion of life to be
found in the jungle could only be supported by rich and fertile soil.

However, the jungle recycles its organic matter so efficiently that it requires few resources from the soil beneath it.

Consequently, those who have destroyed the jungle in order to plant crops, graze cattle or raise specialized tree plantations have soon discovered to their cost that they have violated the true law of the jungle. This kind of dominance leads to extinction for all. After one or two seasons, the nutrients are gone from the soil. The land becomes a desert.

A lion may be 'king of the jungle' by virtue of being its top predator – but he does not, like man, seek to demonstrate that dominance by destroying the jungle itself. The checks and balances of jungle law determine that the prey controls the predator as much as the predator controls the prey.

3

The Morality of Evolution

Evolution: the momentum of life growing and expanding • Anti-Evolution: anti-life forces that diminish and destabilize life • Since the Industrial Revolution • Reversing the process of evolution.

Darwinian theory has been used as an argument for everything from racism to sexism. However, the Social Darwinians have chiefly misused evolutionary theory by insisting that the sacrifice of the organic world to industrial progress is in the natural order of things. To this end, they argue that industrial progress is itself a manifestation of evolution; that industrial man has simply 'speeded up' the process of evolution by accelerating the rate of species extinction.

This idea of seeing a highly increased rate of extinction in terms of accelerated evolution is not just a misinterpretation, it is a concept that is a total travesty of evolutionary theory.

Since the beginning of life on this planet, there has been a steady and continuous growth in the number, diversity and complexity of life-forms. As we have seen in the jungle, life has generated and accelerated the growth of more life. This is evolution.

Since the rise of industrial man, there has been a steady diminishment and elimination of diverse life-forms on a massive scale. This is not accelerated evolution, it is evolution reversed. This is 'anti-evolution'.

These two concepts, in fact, essentially define the two sides in this global conflict of the ecology wars: the pro-life forces of evolution and the anti-life forces of anti-evolution.

It was the British ecologist Edward Goldsmith who first used the term 'anti-evolution' and pointed out that through maximizing genetic diversity and complexity, three billion years of evolution has 'tended in the direction of increased biospheric stability'. Since historic times, however, and particularly since the Industrial Revolution, the biosphere under man's direction has 'tended instead towards decreased stability'.

The evolutionary process created the biosphere. It is a process we are now reversing and we are moving towards increased destabilization.

As Edward O. Wilson, the sociobiologist, once stated: 'The worst thing that can happen – will happen – is not energy depletion, economic collapse, limited nuclear war or conquest of a totalitarian government . . . The one process ongoing in the 1980s that will take millions of years to correct is the loss of genetic and species diversity by destruction of natural habitats. This is the folly our descendents are least likely to forgive us.'

4

Peace

Evolution is not Russian roulette • Evolution teaches
altruism above aggression • Jungle as Eden:
harmony and interdependence • Evolutionary
basis for empathy • The MAD option • Message
from the heart.

So let us now set the record straight. Evolution is not primarily a game
of Russian roulette. The 'law of the jungle' and 'survival of the fittest'
are not savage doctrines after all. Evolution is nothing less than the
force of life diversifying and growing. It is the momentum of life
expanding over the planet.

The moral imperative of evolution points the way for man as it does
for nature. Henryk Skolimowski, in his book *Eco-Philosophy*, makes
exactly this point. 'All those theories of aggression,' Skolimowski
writes, 'which revel in the apparently destructive nature of man and
which are purportedly based on evolution, seem to be quite oblivious
to the work evolution has done through altruism. It is not asserted here
that aggression is not part of our heritage, but only that altruism has
prevailed and will prevail, because it is in the nature of evolution. We
could not live one single day, even in the meanest of societies, without
altruistic behaviour occurring all the time.'

We must acknowledge this fact. Let us join the pro-life forces of
evolution and recognize that the true law of the jungle is that of Eden:
harmony and interdependence, which brings peace. Not a static peace,
but an organic, active, life-enhancing peace. A peace that expands and
stabilizes life's grip on the planet.

We have arrived at a critical moment in human history and the
history of life on earth. The guaranteed dead-end of a nuclear war has
brought the world face to face with the consequences of ignoring the
true law of the jungle. The massive build-up of nuclear arms promising
'Mutually Assured Destruction' – or the MAD option – has long since
put paid to the idea of anyone 'winning' in an all-out nuclear war.

There is only one strategy in the nuclear conflict that will allow
anybody to win. There are finally only two options: either we all

survive, or none of us will survive. We must create a conspiracy by which we may devise that strategy of survival for everyone. Life can only continue on this planet if the people of the world begin to understand the logic behind such a strategy.

This is exactly what Albert Einstein meant when he wrote: 'The unleashing power of the atom has changed everything except our way of thinking . . . we need an essentially new way of thinking if mankind is to survive.' This 'new way of thinking' is the only way to peace in any of the Eco Wars. It does not matter whether the issue is nuclear war, pollution, or extinction of other species. Peace can only be achieved when we understand the wisdom in altruism; when we respond to the prompting of the empathetic heart. We must listen to that small, childlike voice within each of us that tells us truthfully, and without the least doubt: 'There is only one life on this planet, and we are all part of it.'

Bibliography

Abbey, Edward: *The Monkey Wrench Gang* (Avon Books, New York, 1983).

Adamson, Joy: *Born Free* (Chancellor Press, Portland, Oregon, 1986).

Amory, Cleveland: *Man Kind?* (Harper and Row, New York, 1974).

Brookland, John, Hora, Cheryl and Carter, Nick: *Injury, Damage to Health and Cruel Treatment of Live Fauna* (EIA, London, 1985).

Caras, Roger: *Dangerous to Man* (Henry Holt, New York, 1976).

Carson, Rachel: *Silent Spring* (Houghton Mifflin, Boston, 1963).

Carter, Nick and Currey, Dave: *The Trade in Live Wildlife* (Environmental Investigation Agency, London, 1987).

Carter, Nick and Thornton, Allan: *Pirate Whaling 1985* (Environmental Investigation Agency, London, 1985).

Crail, Ted: *Apetalk and Whalespeak: The Quest for Interspecies Communication* (Contemporary Books, Chicago, 1983).

Crosby, Alfred: *Columbian Exchange: the Biological and Cultural Consequences of 1492* (Greenwood, Westport, Ct, 1972).

Day, David: *The Doomsday Book of Animals* (Ebury Press, London, 1981).

—— *The Whale War* (Routledge, London, 1987).

Del Tredici, Robert: *At Work in the Fields of the Bomb* (Harrap, London, 1987).

Domalain, Jean-Yves: *The Animal Connection: The Confessions of an Ex-Wild Animal Trafficker* (Beckman, New York, 1978).

Douglas, Joseph: *CBW: The Poor Man's Atomic Bomb* (Macmillan, London, 1984).

Douglas-Hamilton, Iain and Oria: *Among the Elephants* (Collins, London, 1975).

Ehrlich, Paul and Anne: *Extinction* (Gollancz, London, 1982).

Epstein, S., Brown, L., and Pope, C.: *Hazardous Waste in America* (Sierra Club, San Francisco, 1982).

Foreman, Dave: *Ecodefense: A Field Guide to Monkey Wrenching* (Earth First, Tucson, Arizona, 1987).

The Global 200 Report to the President (Penguin, New York, 1982).

Goldsmith, E. and Hildyard, N.: *The Earth Report* (Mitchell Beasley, London, 1988).

Greenway, J.C. Jr: *Extinct and Vanishing Birds of the World* (Dover, New York, 1967).

Harper, F.: *Extinct and Vanishing Mammals of the Old World* (International Wildlife Protection Publication No.12, 1945, New York).

Harris, Robert and Paxman, Jeremy: *A Higher Form of Killing: Secret Story of Gas and Germ Warfare* (Chatto & Windus, London, 1982).

Hunter, Robert: *The Greenpeace Chronicle* (Picador, London, 1980).

IUCN (International Union for Conservation of Nature and Natural Resources): *Derniers Refuges* (Atlas) Brussels: Elsevier, 1956).

IUCN: *Red Data Books* (SCMU (Species Conservation Monitoring Unit), Cambridge, 1966–82).

IUCN: *Red Data List of Threatened Animals* (SCMU, Cambridge, 1988).

Jordan, Bill and Ormrod, Stefan: *The Last Wild Beast Show* (Constable, London, 1978).

Kidron, M. and Segal, R.: *The State of the World Atlas* (Pan, London, 1981).

Kidron, M. and Smith, D.: *The War Atlas* (Pan, London, 1983).

Komarov, Boris: *Destruction of Nature in the Soviet Union* (M.E. Sharpe, New York, 1977).

Lilly, John: *Communications Between Man and Dolphin: The Possibility of Talking with Other Species* (Crown, New York, 1961).

London, Jack: *Call of the Wild* (Puffin, London, 1983).

May, John and Martin, Michael: *The Book of Beasts* (Hamlyn, London, 1982).

Mowat, Farley: *Virunga* (McClelland and Stewart, Toronto, 1987).

Myers, Norman: *The Gaia Atlas of Planet Management* (Pan, London, 1985).

—— *The Sinking Ark* (Pergamon Press, Oxford, 1979).

Natural Disasters: Acts of God or Acts of Man? (Earthscan, London, 1984).

Nilsson, Greta: *The Endangered Species Handbook* (Animal Welfare Institute, Washington DC, 1983).

—— *Facts About Furs* (Animal Welfare Institute, Washington DC, 1981).

Paterson, Francine and Lindon, Eugene: *The Education of Koko* (Henry Holt, New York, 1981).

Pratt, Dallas: *Alternatives to Pain in Experiments on Animals* (Argus Archives, New York, 1984).

Rashke, Richard: *The Killing of Karen Silkwood* (Penguin, New York, 1981).

Registein, Lewis: *America the Poisoned* (Capitol, Washington, 1983).

—— *The Politics of Extinction* (Macmillan, New York, 1975).

Sagan, Carl: *The Dragons of Eden* (Random House, New York, 1977).

St Barbe Baker, Richard: *I Planted Trees* (Lutterworth, London, 1945).

Salt, Henry: *Animal Rights: Considered in Relation to Social Progress* (Centaur Press, Arundel, 1980).

Salvadori, F.B.: *Wildlife in Peril* (David and Charles, London, 1978).

Sewell, Anna: *Black Beauty* (J.M. Dent (Children's Illustrated Classics), London, 1950).

Shears, R. and Gidley, I.: *The Rainbow Warrior Affair* (Allen and Unwin, London, 1986).

Singer, Peter: *Animal Liberation* (Thorsons, Wellingborough, 1983).

Skolimowski, Henry: *Eco-Philosophy* (Marion Boyars, Boston, 1981).

Stanley, H.M.: *In Darkest Africa* (Greenwood Press, London, 1978).

Stonehouse, Bernard: *Saving the Animals* (Weidenfeld & Nicolson, London, 1981).

US Congressional Hearings: *Review of Global Environment Ten Years After Stockholm* (US Government Printing, Washington, 1982).

Wasserman, H. and Solomon, N.: *Killing Our Own* (Dell, New York, 1982).

Watson, Lyall: *Supernature* (Corgi, London, 1981).

Watson, Paul: *Sea Shepherd* (Norton, New York, 1982).

Weir, David: *The Bophal Syndrome* (Earthscan, London, 1987).

Wells, H.G.: *War of the Worlds* (Pan Books, London, 1975).

Wendt, Herbert: *Out of Noah's Ark* (Weidenfeld & Nicolson, London, 1956).

Wood, Gerald: *The Guinness Book of Animal Facts and Feats* (Guinness, London, 1976).

Ziswiller, U.: *Extinct and Vanishing Animals* (Longman Green, London, 1967).

Index